Philosophical and political struggles against the binary oppositions and hierarchies that have shaped our contemporary globalised world have often reinstated these binaries in their attempts to oppose them. As Doerthe Rosenow shows, the continuation of this modernist thinking within critical thought and politics points to a need to confront colonialist logics that remain wrapped up in not only our history but also our epistemology and ontology. In this compelling study, Rosenow draws upon Bruno Latour, Gilles Deleuze and Félix Guattari, Gyatri Spivak, María Lugones and various other post-structuralist, post-colonial and de-colonial writers to engage critically with environmental activism, and particularly anti-GMO movements. Her complex approach refuses the false alternatives of a simple valorisation or rejection of science, tradition, nature or the perspective of the oppressed, and aims instead at formulating the principles of another, non-binary kind of politics. Proceeding in this way, she refuses to limit herself to Eurocentric critiques of modernism and colonialism, instead pressing herself and her readers to enter into dialogue with what has remained 'Other' to Western thinking. This book will be of interest to scholars and students who are acquainted with established critiques of identity, power, representation, knowledge and resistance, and looking for creative ways to push them further.

Nathan Widder, *Professor of Political Theory,*
Royal Holloway University of London

The production of genetically modified species or organisms (GMOs) is but the latest in the now globally manifest assault on the earth and its environment. For Doerthe Rosenow, author of this highly original and indeed controversial book, a statement such as this would constitute modernist thinking, one that is imbued with a binary logic that perpetuates the assault rather than tackling its misplaced ontologies and epistemologies. In an intellectual trajectory that takes her from Deleuze to Latour, from new materialism to decolonial thought, Rosenow challenges the persistence of modernist and colonial approaches to the environment, unravelling as she does the potentiality of modes of indigenous activism that transcend the dualisms informing contemporary environmental politics. The book should be included on any interdisciplinary syllabus related to critical social and political thought in general and environmental politics, activism, and resistance in particular.

Vivienne Jabri, *Professor of International Politics,*
King's College London

In this rich study, Doerthe Rosenow first complicates the anti-GMO movement by drawing upon Bruno Latour to challenge the culture/nature binary informing the movement and then deepens that discussion by exploring imperial practices that seek to impose such crude understandings upon colonial and postcolonial regions. A fascinating transdisciplinary study that will repay close attention.

William E. Connolly, *Krieger-Eisenhower Professor*
at John Hopkins University, author of Facing the
Planetary: Entangled Humanism and the
Politics of Swarming

Doerthe Rosenow's book offers both an insightful study of anti-GMO activism and a programmatic and challenging theoretical intervention. *Un-making Environmental Activism* raises urgent questions about the relation between the study and practice of environmental politics, analyses of coloniality, and the politics of the new materialisms today.

Andrew Barry, *Professor of Human Geography,*
University College London

T0199469

Un-making Environmental Activism

Much environmental activism is caught in a logic that plays science against emotion, objective evidence against partisan aims, and human interest against a nature that has intrinsic value. Radical activists, by contrast, play down the role of science in determining environmental politics, but read their solutions to environmental problems off fixed theories of domination and oppression. Both of these approaches are based in a modern epistemology grounded in the fundamental dichotomy between the human and the natural. This binary has historically come about through the colonial oppression of other, non-Western and often non-binary ways of knowing nature and living in the world. There is an urgent need for a different, decolonised environmental activist strategy that moves away from this epistemology, recognises its colonial heritage and finds a different ground for environmental beliefs and politics. This book analyses the arguments and practices of anti-GMO activists at three different sites – the site of science, the site of the Bt cotton controversy in India, and the site of global environmental protest – to show how we can move beyond modern/colonial binaries. It will do so in dialogue with Gilles Deleuze, Bruno Latour, María Lugones, and Gayatri C. Spivak, as well as a broader range of postcolonial and decolonial bodies of thought.

Doerthe Rosenow is Senior Lecturer in International Relations at Oxford Brookes University. She is interested in the theorisation and analysis of political struggle in relation to understandings of nature, particularly from perspectives that engage notions of materiality and (de-)coloniality. Her research is interdisciplinary, crossing over the boundaries of international relations, political theory, human geography, anthropology, and continental philosophy.

Routledge Research in Place, Space and Politics
Series Edited by Professor Clive Barnett
Professor of Geography and Social Theory, University of Exeter, UK.

This series offers a forum for original and innovative research that explores the changing geographies of political life. The series engages with a series of key debates about innovative political forms and addresses key concepts of political analysis such as scale, territory and public space. It brings into focus emerging interdisciplinary conversations about the spaces through which power is exercised, legitimized and contested. Titles within the series range from empirical investigations to theoretical engagements and authors comprise of scholars working in overlapping fields including political geography, political theory, development studies, political sociology, international relations and urban politics.

For a full list of titles in this series, please visit www.routledge.com/series/PSP

Un-making Environmental Activism

Beyond Modern/Colonial Binaries in the GMO Controversy

Doerthe Rosenow

Routledge
Taylor & Francis Group

LONDON AND NEW YORK

First published 2018 by Routledge

2 Park Square, Milton Park, Abingdon, Oxfordshire OX14 4RN
52 Vanderbilt Avenue, New York, NY 10017

Routledge is an imprint of the Taylor & Francis Group, an informa business

First issued in paperback 2019

British Library Cataloguing-in-Publication Data
A catalogue record for this book is available from the British Library

Library of Congress Cataloging-in-Publication Data
Names: Rosenow, Doerthe, author.
Title: Un-making environmental activism : beyond modern/colonial
    binaries in the GMO controversy / Doerthe Rosenow.
Description: New York : Routledge, 2017. | Includes bibliographical
    references and index.
Identifiers: LCCN 2017036715 | ISBN 9781138652279 (hardback) |
    ISBN 9781315624396 (ebook)
Subjects: LCSH: Transgenic organisms—Political aspects. |
    Environmentalism.
Classification: LCC QH442.6 .R67 2017 | DDC 636.08/21—dc23
LC record available at https://lccn.loc.gov/2017036715

ISBN: 978-1-138-65227-9 (hbk)
ISBN: 978-0-367-87580-0 (pbk)

Typeset in Times New Roman
by Apex CoVantage, LLC

To Joshua and Charlotta

Contents

Preface

A book usually marks the end of a journey and the beginning of something new. This book made quite a few journeys before arriving where it is now, and rather than being the end of a project, it marks a transition. Having started off as a PhD project a decade ago, it began with an interest in rethinking political resistance based on the thought of Michel Foucault. Like many students of critical International Relations, I had become intrigued by Foucault's radical historical investigations of the interrelation of knowledge and power that were related to his refusal to engage in questions of ontology and metaphysics. However, the more I became interested in the significance of general theory and philosophy for political activism, the more I became frustrated about Foucault shying away from acknowledging the philosophical implications and ground of his work (see Coleman/Rosenow in Ansems de Vries et al., 2017; Coleman and Rosenow, 2016, 2017a, 2017b; cf. Coleman, 2015a). As a consequence, the work of Deleuze, who constructs a metaphysics of transformation in order to challenge modern philosophy, started to become more relevant. Deleuze's work continues to be significant for this book, not least because of Deleuze's ability to make his audience *sense* rather than rationally process the value of his work for addressing the constraints of modern thought (cf. Williams, 2005: 15). Last, the more I engaged with political resistance not just in a generic sense, but with environmental activism in particular, Science and Technology Studies (STS) became an interesting resource. I got hooked on Latour's critique of what he defines as the most central dichotomy of modern thought – the one between human and nonhuman, society and nature – and his attempt to overcome it.

In the spring of 2016 the book project seemed finally in place: I had an objective – the overcoming of modern binaries in environmental activism, a coherent theoretical framework for which I drew on both Deleuze and Latour, and an empirical case, namely anti-GMO activism at various sites. Then in June 2016 I was invited to a workshop at the Centre for Global Cooperation in Duisburg, Germany on the topic of *New Materialisms & Decoloniality: A Conversation*. Organised by Pol Bargués-Pedreny and Olivia Rutazibwa, the workshop aimed to bring into critical dialogue New Materialist and decolonial scholars (as I will outline in greater detail in Chapter 1, the term 'New Materialisms' is actually a contested one among those commonly grouped together in its name). Through the texts that we read together and the discussions that we had, I came face to face with the fundamental Eurocentrism of not only much of the New Materialisms more generally, but also my own approach until this point. Though I myself had worked on environmental activism from a *postcolonial* perspective at one point (see Rosenow, 2013), the *decolonial* approach made me aware that the problem of coloniality was not just one of representation, socioeconomic exploitation or political approach 'out there' in the world of policy-making and political practice. Instead, the problem was the coloniality of knowledge (Mignolo, 2007: 451) that characterises the very (European) concepts and categories that I myself had used to make sense of the world, including the concepts of my cherished poststructuralist and STS authors.

It was this workshop that introduced me to Zoe Todd's influential critique of Latour. Recounting her attendance of a talk given by Latour, Todd (2016) reflects on the ongoing Eurocentrism of dominant scholars in critical thought: 'I waited through the whole talk, to hear the Great Latour credit Indigenous thinkers for their millennia of engagement with sentient environments, with cosmologies that enmesh people into complex relationships between themselves and all relations . . . It never came. I was left wondering, when will I hear someone reference Indigenous thinkers in a direct, contemporary and meaningful way in European lecture halls?' (ibid., 5–7) Todd's article in particular confronted me head-on with the biases of my education, no matter how 'critical' I thought my studies had been. All of the theorists that I had seriously engaged to that date were white, French men. My diagnosis of central problems and my suggestions for transformation were all based in a body of thought, concepts and critiques that proceeded from what Arturo Escobar (2007: 180) calls an 'intra-modern' perspective: critiquing modernity from within, with little reference to anything that co-constituted the rise of modernity outside of Europe. My book fell to pieces and my research was in crisis. What was I supposed to do? Start from scratch? I already had a book contract, a deadline. First, I thought that a disclaimer in the Preface might suffice. But when I turned to my Chapters 3 and 4 that still needed re-drafting, I could not but engage with the decolonial critique. Despite this, I decided to be pragmatic and not start completely from scratch, which means that there are some decisive omissions in this book. Most notably, I did not end up addressing Todd's above-cited concerns about the need to substantially engage Indigenous/non-Western thought. Rather, the decolonial ethos of this book manifests itself in the attempt to 'shrink' modernity to one 'ontological enactment' among others, creating space for alternatives to rise (Blaser, 2013: 553). The result is indeed a book in-between: the beginning of a journey rather than an end. It probably does not fully live up to the demand of the academy for stringency, coherence and logic. It also does not do justice to the richness and complexity of the literature that I have only recently started to engage. But maybe these 'failures' make the book truer to its purpose than it appears at a first glance. To say it with María Lugones (2003), rather than featuring a clean and pure identity, the book is 'mestizaje', impure, curdle-separated, kind of both, neither/nor. Whatever it is that came out of this process of shifting my work: it definitely *feels* right.

This book has been profoundly shaped by the intellectual and emotional support, love, friendship and comradeship that I have encountered at every level. My deep gratitude goes first and foremost to my wonderful friend and co-author Lara Coleman. Intellectually, professionally and as a 'whole person', I would not be where I am today without Lara's wisdom, warmth, generosity, intellectual feedback and feminist solidarity. In terms of my conceptual work, she was often the first to come up with decisive ideas, and to get my and our collaborative work to new levels with her famous 'hunches'. This book would not be the same without the helpful feedback of those who have read and commented on very early and final draft chapters: Andrew Barry, Pol Bargués-Pedreny, Gary Browning, Lara Coleman, Martin Coward, Luis Lobo-Guerrero, Tina Managhan, Meera Sabaratnam, Lisa Tilley, William Walters and Nathan Widder. I want to particularly thank Andrew Barry and William Walters, without whose generous and helpful feedback in the official review of the book proposal I would have not been able to finally leave behind my PhD style of writing ('Be bold!' as one of them concluded his review). My particular thanks also go to Gary Browning, for having read multiple drafts of the introduction and conclusion and for providing me with a spot-on one-page summary of the book when I was no longer able to see the forest for the trees. The same goes for Nathan Widder, who has been a brilliant mentor ever since I got to know him at Royal Holloway University of London many years back. Nathan helped me to put the book proposal together and also gave fantastic feedback on the Deleuzean parts towards the end of the project.

A big thank you to my great friends and colleagues at Oxford Brookes University, for their emotional and intellectual support, their trust, and for believing in me: Victoria Browne, Tina Managhan, Chris Hesketh and Maïa Pal. Thanks to my academic friends more generally, for always providing intellectual stimulation and platforms for – often heated! – philosophical and political discussion: Leonie Ansems de Vries, David Chandler, Nicholas Kiersey, Kai Koddenbrock, Nicholas Michelsen, Louiza Odysseos, Chris Rossdale, and Karen Tucker. To my former PhD supervisor Claudia Aradau, who has accompanied this project from the very beginning. Once again thanks to Pol Bargués-Pedreny and Oliva Rutazibwa for having organised the workshop in Duisburg that changed everything, and also to my co-authors of the collective discussion 'Fracturing politics (or, how to avoid the tacit reproduction of modern/colonial ontologies in critical thought)' that was published in volume 11, issue 1 of *International Political Sociology* and introduced me, through Rolando Vázquez' contribution in conversation with Lara Coleman, to decolonial thought: Leonie Ansems de Vries, Lara Coleman, Martina Tazzioli and Rolando Vázquez.

Thanks to my friend Kärg Kama from the University of Oxford who was always able to put things in perspective, and to the Home community in Oxford for spiritual nourishment. To my parents, for providing a safe space for retreat, reflection and unconditional love. Finally: to Jan, who has been with me on every journey for now two decades and who has always loved me no matter what. To my beautiful children Joshua and Charlotta, for being such a joy and for always pointing me back to what really matters in life.

Abbreviations

ANT	Actor-Network-Theory
Bt	bacillus thuringiensis
DNA	deoxyribonucleic acid
EC	European Communities
GE	genetic engineering
GJM	Global Justice Movement
GM	genetically modified
GMO	genetically modified organism
HGP	Human Genome Project
ICC	Intercontinental Caravan
IMF	International Monetary Fund
KFC	Kentucky Fried Chicken
KRRS	Karnataka Rajya Raitha Sanga (Karnataka State Farmers Organisation)
MNC	multinational corporation
NGO	nongovernmental organisation
RNA	ribonucleic acid
WSF	World Social Forum
WTO	World Trade Organisation

1   Un-making environmental activism

Tucked away in the countryside of Northern Germany lies the Wendland, which is a region that to the outsider does not stand out in any particular way. But over the past 40 years the Wendland has become a significant location for German anti-nuclear power activists, after a salt dome in one of its small villages – Gorleben – was chosen to be the *Zwischenlager* (intermediate storage facility) for Germany's nuclear waste in the 1970s. Today the Wendland means a lot to Germans like me who have undergone their initiation into environmental activism by participating in the protests against nuclear waste transports that make their way to Gorleben on a yearly basis. In these protests, Greenpeace activists join anarchists as well as local farmers to occupy rail tracks and roads. In fact, the whole Wendland population participates, with the anti-nuclear symbol 'X' being exhibited in many windows. I still vividly remember my own 2003 Gorleben protest in which I was given a lesson on the merits of the former Socialist German Democratic Republic at the booth of the Marxist-Leninist Party and learned about the dangers of nuclear power at the Greenpeace stall, while my partner – a Greenpeace activist – danced with a fiddler and his children on the occupied rail tracks. At the time, being a German environmentalist for sure meant to be 'anti': anti-nuclear power, anti-biotechnology, anti-rainforest logging and, already looming on the horizon, anti-man-made climate change.

Since that time the once clear waters of my environmental beliefs have become muddied. Environmentalists have started to prominently discuss whether some environmental issues are more significant than others, and whether there are even outright contradictions between the various 'anti'-positions. In the UK, where I have been living at the time of writing for more than a decade, the renewed commitment to nuclear power is (among other things) justified by pointing at the need to protect the climate, based on the argument that the nuclear industry is allegedly low carbon. This is an argument not just made by the UK government (see e.g. Leadsom, 2016), but also by some prominent environmentalists. George Monbiot, for example, argues that the fighting of nuclear power is counterproductive for environmentalists insofar that it distracts them from the problems that really matter, such as the CO_2 emissions of the coal industry (which, Monbiot (2013) suggests, can only be tackled by advocating an energy mix that includes nuclear power). Similarly, some argue that the issue of climate change should make us environmentalists change our attitude towards agricultural bio-technology, the activism around which lies at the heart of this book. Environmentalist and popular writer Mark Lynas, for example, publicly 'converted' to a pro-genetically modified organisms (GMO) position based on the argument that he has to be consistently 'pro'- or 'anti'-science. He cannot, so his reasoning, be 'anti-science' in relation to agricultural biotechnology, whilst being 'pro-science' in relation to climate change (Lynas, 2015). There is, he (ibid.) maintains, a sci-entific consensus for both the actuality of man-made climate change and the safety of GMOs.

What are the reasons for this professed desire to organise into a hierarchy and/or play against each other environmental beliefs, with agricultural biotechnology and nuclear power being two of the most prominent issues losing support among at least some environmentalists?[1] As Lynas's comments make clear, the distinc-tion between 'proper' and 'improper' environmental beliefs is often grounded in science as the decisive site for verifying claims about how nature needs to be protected best. Indeed, those environmentalists who continue to be concerned about agricultural biotechnology often counter pro-biotech arguments with the same logic: they point out that the 'evidence' for a pro-position is either not there or that it is inconclusive; in other words, they contest the idea that there really is a scientific consensus. This is often related to the allegation that those in favour of biotechnology are not independent scientists, but are compromised by their ties to industry (see e.g. Robinson, 2015). In fact, as I was writing the first version of this introduction, Riverford veg box scheme owner Guy Watson responded in the newsletter that accompanied my box for that week to a pro-GMO BBC *Panorama* episode that had just been aired in the following way (Watson, 2015; emphasis added):

> I remain open-minded about the benefits that GM might bring in the future . . . but . . . bombarding us with *emotive messages* driven more by a PR agenda than by *fact* is unforgivable. We need, rather, a *cool headed evaluation of the scientific evidence and the commercial interests at play.*

One of the most interesting features of this quote is the binaries that it invokes, which also (at least implicitly) structure the arguments of those in favour of biotech: the 'cool' scientific versus the 'emotive' other, 'pro' versus 'anti'-science, (objective) 'evidence' versus (commercial or otherwise partisan) 'interests'.

In this book I will argue that this urge towards binarisation is grounded in a taken-for-granted modern epistemology that is delimiting our understanding of nature, reality and political transformation. Crucially, this epistemology has his-torically come about through oppressing and annihilating other, non-Western and often non-binary ways of knowing nature and living in the world, which is why I call previously mentioned binaries not just *modern*, but also *colonial*. What interests me in this book is what sort of environmental politics and activist strat-egy could emerge if we moved away from this epistemology and dropped this urge towards binary categorisation, if we acknowledged its colonial heritage and (consequently) found a different ground for environmental beliefs, values and politics more generally (cf. Braun, 2002)?

There are of course other, more radical forms and arguments of/in environmental activism. Indeed, the approach outlined so far could be called a mainstream, 'pacified', evidence-based one that usually complies with paradigms such as ecological modernisation, sustainable development and green growth. Such an approach sees effective governmental regulation as the best response to enduring environmental problems, instead of trying to change the fundamental oppressive socioeconomic structure of society. This institutional approach to 'saving' the environment has always stood in 'historical tension' with more radical environmental Marxist or anarcho-autonomist branches (Reitan and Gibson, 2012: 396–7). However, one of the contributions of this book lies in the argument that it is not only mainstream environmentalists that sign up to a modern/colonial epistemology, but also, as I will outline in one of the following sections of this introduction, many radical activists and scholars. Drawing on decolonial thought (e.g. Quijano, 2007; Mignolo, 2000), I will argue that radicals, while acknowledging that other, non-Western ways of making sense of nature have been ignored and suppressed in *political and socioeconomic practice*, often neglect the way that Marxist, anarchist, deconstructivist or otherwise radical Eurocentric *concepts and categories* continue to suppress alternative bodies of knowledge about the world.

This book engages with a particular environmental issue and the activism against it: agricultural biotechnology/GMOs. This issue is interesting because two of the rationales that underlie activists' fight against GMOs have in themselves the potential to disrupt modern/colonial binaries. First, anti-GMO activists who continue to argue against biotechnology do so, as Lynas rightly points out, *against* the scientific consensus. However, that does not mean that they altogether abandon the scientific argument: drawing on complexity science they maintain, against mainstream molecular biology, that the organism (and/or nature as such) is a complex, self-coordinating entity that cannot be externally controlled. As I will show in Chapter 2, they contest the traditional binary between the scientist as subject and the organism as object and instead depict observer and observed as intermingled. With this they at least implicitly go against the crucial modern mind/matter dichotomy as well as related understandings of cause and effect working in a linear manner, and the possibility of predictability and control. Some anti-GMO activists see the latter as being the outcome of a human chauvinistic attitude towards nature that needs to be overcome. At least at a first glance, this argument seems to be close to many non-Western cosmologies in which the human and the nonhuman have never been as clearly distinguishable as they have been for the modern subject. However, as I will also show in Chapter 2, when it comes to the concrete political argument that activists make about GMOs, modern/colonial binaries re-enter the picture, particularly when a strong distinction is made between the 'natural' organism and the 'unnatural' GMO. Paradoxically, this distinction is based on the assumption that the GMO is a bounded entity the identity of which can yet again be described on the basis of inherent, fixed and stable properties (rendering it 'unnatural'). This too easily leads to a call for excluding and destroying the 'unnatural' or 'monstrous', as well as a collapsing of the true and good into the 'natural'. As I will show particularly in Chapter 4, both moves have very problematic implications for the way that non-Western cultures, societies and bodies of knowledge are perceived, judged, appreciated and/or excluded.

The second rationale of anti-GMO argument and practice that makes it interesting for challenging and overcoming modern/colonial binaries in environmental activism explicitly builds on a critique of neocolonial/neoliberal structures of domination. Prominent Indian intellectual Vandana Shiva, for example, calls Western environmental science 'masculine', instrumental and exploitative (see e.g. Shiva, 1989). Shiva makes a link between this understanding and the developmental and economic agenda advanced by multinational corporations (MNCs) and Western-led international organisations in relation to agriculture (see e.g. Shiva, Emani and Jafri, 1999). Shiva argues that this Western understanding should be replaced by an approach towards agriculture and development that draws on non-Western, Indigenous, nurturing ways of engaging with land and nature. As I will show in Chapter 3, agriculture is indeed a prime site for making visible what some scholars have called 'ontological incompatibility' (Carro-Ripalda and Astier, 2014). This means that the modern understanding of what it means to be human, of how the human relates to the nonhuman, and what the general place of humans and nonhumans is in the cosmos (all of which is significant for doing agriculture) not only clashes with non-modern understandings, but is incomprehensible to the latter (and vice versa).[2] However, based on an analysis of the Indian controversy around Bt cotton, I will argue in Chapter 3 that the continuous invocation of the 'external' (Western states, Western ways of doing agriculture, MNCs, international organisations) versus the 'internal' (traditional, Indigenous, democratic, local ways of doing agriculture) reinforces modern/colonial binaries at the same time as it challenges them. As the case of the Indian Bt cotton controversy manifests, the distinction that anti-GMO activists make between the 'natural' and the 'unnatural' does not only refer to the GMO as opposed to the 'natural' organism, but also to agriculture as a whole: the 'unnatural' is linked to the industrial approach, which follows the ideas of the Green Revolution, whereas the 'natural' is embodied in the traditional, authentic, Indigenous approach. As I will argue, anti-GMO activists need to understand and challenge modernity/coloniality at a much deeper epistemological and ontological level; tracing it back to particular colonial and postcolonial trajectories. In the case of Indian agriculture, for example, activists need to understand how colonial approaches towards state and economic development that have wholesale swallowed colonial ontologies and rationales make it very difficult – if not even impossible – to hark back to an 'authentically' different ontology and agricultural practice. As I will show in Chapter 4, the continuous upholding of the modern=bad/traditional =good binary also has profound implications for the activist attempt to join forces across diverse geographical, cultural and socioeconomic backgrounds – i.e. the attempt to make environmental protest *global* – which too easily leads to a reproduction of colonial oppression through reading protest rationales, strategies and activist identities through one particular (modern) epistemic frame.

In sum, this book studies the problem of modern/colonial binaries in relation to anti-GMO activism at altogether three sites: the site of science (Chapter 2), the site of India's controversy around Bt cotton (Chapter 3) and the site of global environmental protest (Chapter 4). It does so by analysing the activists' arguments and practices that we find at these sites, as well as by using certain conceptual and ontological resources to both understand and move beyond the taken-for-granted (colonial) 'common sense' of existing positions. The resources that I will use are in particular Bruno Latour's (2004) attempt to overcome the modern society/nature binary by assembling humans and nonhumans in a new political collective, Sylvia Wynter's (2003) foundational essay on coloniality, Gayatri C. Spivak's (1988) famous questioning of the subaltern's voice and her general take on power and representation, Gilles Deleuze's non-binary metaphysics of transformation, and María Lugones's decolonial approach. Some of what I have described above, particularly in relation to the Indian Bt cotton controversy, sounds close to a traditional *postcolonial* approach, insofar that I question how colonialism has in itself created understandings of the 'natural', the 'traditional' and the 'authentic' from which the 'postcolonial' is unable to get away. Relating this interrogation to the question of how to pursue a *postcolonial* environmentalism brings this book close to Bruce Braun's work on conceptualisations of and activism around the temperate rainforest in British Columbia. In his seminal book *The Intemperate Rainforest: Nature, Culture and Power on Canada's West Coast* (2002) Braun aims 'to strive towards a new set of concepts that might inform a radical environmentalism that is attuned . . . to the relations of power and domination that infuse our environmental ideas and imaginations', with particular attention being paid 'to the subjugated histories and buried epistemologies – often *colonial* epistemologies – that are hidden by, or within, the terms and identities through which forest politic . . . is organized and understood' (ibid., x; emphasis in original). However, in addition, the aim of my book is a *decolonial* one. Indeed, I aim to use some of the mentioned conceptual and ontological resources to not just analyse the debates, but to pinpoint how the exclusion of non-modern ontologies and epistemologies that has become cemented in much anti-GMO argument and practice has been central to the rise of the modern project as such. In other words, I aim to show how modernity and coloniality have to be understood as co-constituted (Quijano, 2007).

Due to its transitional nature (cf. Preface), the book does not substantially engage with non-modern bodies of thought and practice, as the decolonial approach encourages us to do. However, it aims to use previously mentioned conceptual and ontological resources in dialogue with anti-GMO rationales and arguments to open up space for *sensing* the existence of ontological difference and incompatibility. As I will show throughout the chapters to come, this will make the book in itself become a means of disrupting modern/colonial binaries. Finally, the book will attempt to provide practical suggestions for how to concretely move anti-GMO activism beyond modern/colonial binaries. As the subsequent chapters and particularly the conclusion of this book will show, these suggestions are often (though not always) counter-intuitive to common environmentalist belief.

For example, I will make a case for the need to respect the rights of GMOs for ontological self-definition, the need to let go of an 'anti'-attitude at specific sites, and the need to understand and come to terms with the 'unnatural' monstrosity of nature.

Anti-GMO activism past and present

About 12 percent of arable land worldwide is currently cultivated with GM crops. In 2013, 27 countries used the technology of genetic engineering (GE) in their agricultural production, though most GMOs were grown in just five: the US, Brazil, Argentina, Canada and India (Macnaghten, Carro-Ripalda and Burity, 2015: 8–9). Biotechnological research and its application had begun in the US in the 1970s and had been initially accompanied by US public concern. But this concern waned after scientists had managed to convince the US public that the risks were both 'marginal and manageable' if certain guidelines were followed (Torgersen et al., 2002: 35). By contrast, due to strong public opposition, the Western European market has remained virtually closed to GM-products until very recently (Schurman and Munro, 2009: 156). Established NGOs such as Greenpeace and Friends of the Earth had taken up the fight against GM technology, as had more radical organisations and social movements such as 'Gendreck Weg!' or 'Earth First!'.

In the so-called Global South there has been a rapid growth in the planting and selling of GMOs in the last decade. In 2011 more land was cultivated with GMO crops in the Global South than in the Global North (Macnaghten, Carro-Ripalda and Burity, 2015: 9). While the US, Canadian and Argentinian markets have become saturated, Brazil and India (as 'late adopters') have continued to expand (ibid.). However, both Brazil and India have also featured some of the strongest farmer-based anti-GMO movements in the world. In the late 1990s and early 2000s campaigns against transnational agribusiness in the Global South were coordinated by the People's Global Action network, which included hundreds of Indian farmers and members of the Landless Movement of Brazil (Kousis, 2010: 230). While worries about safety have for a long time constituted the core of concerns in the Western world – particularly in Scandinavian and German-speaking countries, which became the stronghold of European anti-GMO activism in the late 1990s and early 2000s (Torgersen et al., 2002: 60) – campaigns in the Global South have been dominated by issues such as corporate control over seeds and the neoliberal context of the implementation of the technology (Kinchy, 2012: 14; cf. Schurman and Munro, 2010: 57; Jasanoff, 2005: 38).

Although contemporary expressions of protest around agricultural biotechnology remain in many ways wedded to earlier arguments (Kinchy, 2012: 14), more recently the rhetoric has globally shifted towards a suggested need to maintain the right to choose which way to farm, and the socioeconomic consequences that infringing on those rights might entail (ibid., 133; cf. Alessandrini, 2010: 9; Kousis, 2010: 236). In March 2014, for example, Mexico rendered a verdict against agricultural giant Monsanto banning the planting of GE maize in the state of Campeche on the grounds of protecting local Indigenous communities. Growing

GMOs, it was argued, would go against 'the local communities' right to decide on what grows on their land' (Dîaz Pérez, 2014).

The debate on rights is closely related to discussions about the possibility of the coexistence of GM- and other 'natural' or 'organic' products, and the potential danger of the former 'contaminating' the latter. In the European Communities (EC) the issue of coexistence was already debated in the mid-2000s and resulting regulations, as well as those that make the labelling of GM-products compulsory, have led to a considerable dying down of much anti-GMO sentiment (Jasanoff, 2005: 144–5). This went hand in hand with a significant recovery of the biotech industry in the late 2000s, when it was able to capitalise on the spike in global food prices in 2008 and a growing demand for biofuels due to increasing concerns about climate change (Schurman and Munro, 2010: xiv). In North America, by contrast, the lack of formal coexistence rules and the fact that the liability of ensuring that agricultural products remain GM-free lies on the shoulders of individual farmers have led to an increase in protests in the last decade (Kinchy, 2012: 130–1). This had led some researchers, such as Rachel Schurman and William Munro (2010: 180–2), to argue that simple 'pro'- and 'anti'-GMO arguments have started to lose ground. However, resistance to 'the GMO', including calls for its ban or, at least, tight regulation continue to remain central to activism over agricultural biotechnology – exemplified by the neat division that is made between GMOs and 'natural' products, and expressed in the idea of 'choice' (cf. Ansems de Vries and Rosenow, 2015: 1122).

The 'radical' argument against science-based environmentalism

As the previous section has made clear, much of the GMO controversy in Europe has taken place around the question whether or not science has established that GMOs are 'safe' to plant and consume. This focus on science and scientific evidence is typical for Western, modern societies and it has long been questioned by critical theorists. Michel Foucault is one of the prominent thinkers coming to mind when reflecting on how the emergence of 'true' (scientific) knowledge is always interrelated with a particular constellation of political, social and economic power relations. Foucault deemed the human sciences such as psychology to be the decisive sites for the production of acceptable knowledge in modern societies – not because of their objective truthfulness, determined by some 'internal dynamic' (cf. Fuller, 2007: 3), but because they allow an existing set of power relations to function (Foucault, 1991: 27).

Drawing on this and similar accounts, many contemporary critical scholars interested in environmental politics caution us against arguments that rely on 'hard science' as the decisive site of veridiction, because these arguments marginalise those voices that link the need for environmental protection to the need for a more radical economic and political restructuring of society. As some scholars (e.g. de Goede and Randalls, 2009; Swyngedouw, 2007) point out, radical political approaches (rightly) focus on problems that are internal to society, such as structural inequality and exploitation. By contrast, a science-based approach has a tendency to advocate technocratic regulatory solutions for environmental problems, as the latter are regarded as existing externally to the way contemporary society is structured. Here 'nature' is understood as a given object, the comprehension of which relies on technical, calculable knowledge, graspable with numerical evidence, eschewing the 'political' question about 'what kinds of natures we wish to inhabit, what kinds of natures we wish to preserve, to make, or, if need be, to wipe off the surface of the planet' (Swyngedouw, 2007: 23). For Eric Swyngedouw (ibid., 21) the debate over GMOs shows particularly well how one particular kind of 'Nature' (the one that can be 'packaged, numbered, calculated, coded, modelled, and represented') has become hegemonic, and is used by technocrats, scientists as well as environmental activists to ultimately (possibly unintentionally) fend off more fundamental challenges to our world. What underlies Swyngedouw's critique is an understanding of 'politics' as the struggle of different visions about how to structure society as a whole, while 'post-politics' makes political procedure a consensual exercise that erases any notion of fundamental struggle (cf. Rancière, 1999, 2010; Žižek, 1999).

Others argue that scientific evidence allows neoliberal ideologies an ever-tighter grip on contemporary societies. Inspired by Foucault's notion of 'biopolitics', which Foucault uses to describe the intimate relationship between the governance of populations and the knowledge about their presumably 'natural' processes that is available to those in power, it is maintained that neoliberal regimes make use of knowledge to be found in e.g. systems ecology. Based on this knowledge, neoliberals, the argument goes, make a case for the benefits of self-regulation and the becoming 'resilient' of socio-ecological systems (see e.g. Walker and Cooper, 2011). Emphasising the 'complexity' of these systems, neoliberal theorists and policy-makers supposedly maintain that the idea of controlling socio-ecological systems needs to be replaced with a focus on adaptation, so that external catastrophes such as climate change or terrorism can be survived. Elaborating on this logic, Julian Reid (2012: 69) argues that environmentalists consequently need to give up the idea of 'ecological reasoning' as means of political contestation. Instead they need to recognise that the decisive problem of our time is not that there is insufficient concern for the vulnerability of the natural world in mainstream policy-making. Indeed, there is rather too much: neoliberal governance today is, Reid (ibid.) suggests, all about our vulnerability to nature and other external forces (cf. Chandler, 2012; Evans and Reid, 2013, 2014). Instead of misrecognising how ecological systems work, Reid (2012: 68–9, 77) maintains that contemporary neoliberal governmentality has understood this far too well: it has made it part and parcel of the neoliberal logic of self-reliance and responsibility.

All of these analyses and arguments allow for insight into the problems that occur when 'science' is invoked in order to make a case for political action on behalf of some diffuse entity called 'Nature' or 'Ecology'. Particularly the critique of the post-political also rightly emphasises the need to listen to marginalised voices that connect environmentalism to a politics that aims to challenge structures of injustice at a deeper level. However, I am concerned about the extent to which critique in these approaches remains wedded to what Walter D. Mignolo, drawing on Aníbal Quijano, has called the coloniality of knowledge (Mignolo, 2007: 451; cf. Quijano, 2007). The theoretical framework that is used remains an intra-modern one that is sitting, unchallengeable, on top of the critique that is made. Indeed, both the notion of 'post-politics' and the notion of 'bio-politics' have been elevated to the status of 'fixed' theoretical schemata against which existing political practices (such as environmentalist ones) are judged in a binary manner (cf. Coleman and Hughes, 2015: 142; Coleman and Rosenow, 2017a). The 'true' problem that needs to be tackled, the argument goes, is either the socioeconomic structure of Western societies, or global neoliberal biopolitics, both of which being exclusively analysed through the lens of given (European) categories and concepts. In those conceptions 'nature' as something 'out there' is nothing but a construction, a discourse that serves capitalist, neoliberal elites. The implications of this are particularly stark when David Chandler and Julian Reid (forthcoming) use their understanding of the nature of contemporary neoliberal governmentality to explicitly criticise the call for challenging modern/colonial epistemologies and for turning to 'outside' ontologies such as the ones found in Indigenous thought and practice. Instead of seeing those arguments as radical or emancipatory, we should, Chandler and Reid maintain, see them as extending 'dominant western hegemonic practices' (i.e. neoliberal governmentality) (ibid.; emphasis added). Though we should stand in 'political solidarity' with Indig-enous struggle e.g. against the 'repossession of the lands of which it has been robbed', any attempt to use Indigenous knowledge in order to understand 'politics and ontology' anew 'through a way of being' leads us into celebrating 'defeat' rather than 'resistance' (ibid.). Here we can precisely see the problem of a 'radi-cal' scholarship that affirms socioeconomic (decolonial) struggles, but fails to come to terms with the epistemological and ontological dimension of coloniality. The decisive problems of our world are solely grasped through the lens of given (Eurocentric) concepts, while Indigenous cosmologies are reduced to mere 'dis-courses' that are appropriated by dominant governmental regimes and logics. The eye of the critic remains firmly fixed on the coloniser and what He does, with no scope left for 'outside' knowledge to regain any sort of agency.[3] The coloniality of knowledge remains unchallenged.

Moving beyond modern/colonial binaries? The New Materialisms and Latour's politics of the collective

Compared to the arguments that I have outlined and challenged in the previous section, the critical tradition of the so-called New Materialisms could be con-sidered as having a natural affinity with the decolonial project of overcoming the coloniality of knowledge. This is because it explicitly engages in ontological re-imagination in order to critique and potentially overcome modern conceptions of nature, matter, agency and politics. Admittedly the label 'New Materialisms' is in many ways misleading because it tends to conceal the fundamental disagree-ments between the scholars who are grouped together in its name. Here, I use the label as it is employed and explained by Diana Coole and Samantha Frost in their edited volume *New Materialisms* (2010b), and in relation to the scholars who have contributed to that volume. Many of the latter share an interest in developing different understandings of materiality in order to fundamentally challenge modern philosophical frames.[4] Even more importantly, there are quite a few similarities between the 'new' ontologies outlined by those scholars and the understandings of those anti-GMO activists who draw on complexity theory in order to make their case against the technology of genetic engineering (GE) (cf. Braun, 2015).

According to Coole and Frost (2010a) New Materialists attempt to altogether break with binary understandings of the world. Individual entities (human as well as nonhuman) are regarded as fundamentally interconnected and (thereby) unable to hold fixed, contained identities. For William Connolly, for example, the world is one of 'abundance' in which individuals are 'not exhausted' by their identity, but relate to wider 'force-fields' on which they emerge, fracture and (re-)connect (Connolly, 2002: 120, 2011: 5). The force-fields that condition different identities are fundamentally non-binary, and any identity emerging from them is never sufficiently 'finished' to be constituted in opposition to something else. In line with this, any interpretation of reality, of its truth and of its problems, is always exceeded by the ontology it tries to capture, which disrupts any sense of certainty about oneself, the world and the problems that we face. It disrupts any sense of 'attunement, explanation, prediction, mastery, or control' (Connolly, 2011: 5; cf. White, 2000: 114). All of what is usually assumed to be binary becomes part of a non-binary, vibrant, agentic process of materialisation. Politically this should encourage us, as Connolly argues in his latest work which is particularly concerned with the problem of climate change, to form 'a new pluralist assemblage organized by multiple minorities drawn from different regions, classes, creeds, age cohorts, sexualities, and states'; engaging in a 'politics of swarming' that is 'activated by the power of planetary forces' (Connolly, 2017: 8–9).

However, if agency is everywhere and if materiality is vibrant, it becomes difficult to recognise an actual *lack* of it on the side of those who are powerless and oppressed. 'Lack' points to a fundamental negativity that is difficult to integrate into an ontology of 'abundance'. If 'structures' are replaced with 'networks' (as happening for example in Actor-Network-Theory (ANT)), 'all parties are presented as exerting their own kind of power over each other, according to alliances they can form in a given circumstance' (Fuller, 2010: 11). In other words, everything seems to be constantly in flux and open for new events and alliances to take shape. It is often critiqued that New Materialist ontologies occlude how the interrelation of (material) identities is impacted on by particular historical 'power geometries' that have led (and continue to lead) to agents being 'affected' differently (Tolia-Kelly, 2006: 213–14). In other words, structural inequalities and oppression (in their historical and contemporary manifestations) have had an impact on how different agents are affected and how they connect – or indeed fail to do so (ibid.; cf. Ansems de Vries and Rosenow, 2015: 1120). As Gayatri C. Spivak (1988) poignantly argues in her famous essay 'Can the Subaltern Speak?', the problem of dismissing representation is that it erases any possibility to reflect on differentiated positions, and on the fact that people, for historical and geopolitical reasons, have differentiated capacities to act and speak. Pulling everyone and everything together as emerging from some infinite force-field harbours the danger of not being able to interrogate how particular identities emerge as differentiated, shaped by particular sedimented historical (colonial) legacies. Even if the injustices of colonial and neoliberal capitalist exploitation are acknowledged and analysed, as in Connolly's case (see e.g. Connolly, 2008), the call for a 'politics of swarming' that 'is composed of multiple constituencies, regions, levels, processes of communication, and modes of action' (Connolly, 2017: 125) makes it difficult to reflect on varying distributions of affect and connectivity *within* the assemblage. What are the internal hierarchies and structurally differentiated positions, what do they tell us about past and contemporary domination, and what are, as result, the specific responsibilities of those participating in the assemblage?[5]

As this book will show in relation to biotechnology, much of conventional science (in practice and theory), still heavily relies on Cartesian subject-object or mind-matter dualisms: in the process of scientific investigation, the truth of the object is revealed to the scientist as subject, contributing to scientific progress and delineating a world in which those who have minds are agents, and in which matter stays passive and predictable. In other words, GMOs are precisely *not* regarded as agents that exercise an (not entirely predictable) 'force' (cf. Barry, 2013: 2; Whatmore, 2009: 588; Whatmore, 2002), but as objects that are controllable by the human agents who have modified them. By contrast, for Isabelle Stengers, GMOs are nonhuman agents that are brought into the political realm by humans (e.g. anti-GMO activists) who act as their 'spokespersons' and who thereby make the organisms' 'presence felt' (Stengers, 2010: 20).[6] It is argued that this human-nonhuman-assemblage leads to an extension of debates about expertise, a questioning of the limits of knowledge and a contestation over what GMOs really are: objects of progress that are under control, or (for example) 'vehicles for intellectual property rights' that illustrate 'capitalist expropriation strateg[ies]'? (ibid., 21) However, Stengers also deliberately takes the 'oppositional connotation' out of Deleuze and Guattari's concept of the 'minoritarian', and puts it into a context of 'together-ness', where people and things are 'attached' and 'associated' in heterogeneity (ibid., 14). They then have the 'capacity . . . to band together and act in concert but in the manner of a "swarm" rather than in consequence of some pre-figured category of political interest . . . or class' (Whatmore, 2009: 592). But isn't it precisely the oppositional nature of the concept of the 'minoritarian' – or of a concept such as 'class' – that makes it politically powerful, able to disturb the given social and political order by pointing at fundamental structures of oppression and injustice?

As I will show in the following, this problem is at least partly addressed in the work of what Graham Harman (2014) calls the 'middle Latour'. For Latour, the binary between 'nature' and 'society' is decisive for the organisation of modern life: it has developed, he (2004: 3) suggests, 'over centuries in such a way as to make any juxtaposition, any synthesis, any combination of the two terms impossible', and it stabilises all other traditional modern dichotomies: man/

nature, subject/object, mind/matter, etc. With regard to the political implications of his work, Latour has moved, Harman (ibid.) suggests, through different phases. In his early work, Latour is most inspired by the political philosophies of Thomas Hobbes and Niccolò Machiavelli, though he replaces the struggle between *human* power players that we find in both with the struggle of *human and nonhuman* 'actants' the agency of which is solely measured in line with the effect that they are able to generate. Latour's early work is characterised by a '"symmetrical" ontology' in which actants are seeking allies to connect to in order to establish hegemony (ibid., 51). '[M]ight' trumps 'right' (ibid., 13), with 'no court of appeal beyond winning or losing' (ibid., 46). By contrast, the 'middle Latour' replaces this power struggle with a 'carefully assembled institutional network' (ibid., 56). Particularly in his book *Politics of Nature: How to Bring the Sciences into Democracy* (2004), Latour argues that we need to create a 'new Constitution' and set up a new 'collective' that very well serves as a 'court of appeal' for the various propositions that plea for inclusion. He explicitly attempts to establish a practical 'procedure' for making things 'right', one that leaves behind the 'old', modern, binary Constitution. It is for that reason that this part of his work will be one of the crucial conceptual resources that will be used in this book.

In *Politics of Nature* Latour uses the metaphor of the 'house' to describe the make-up of what he calls the 'old [modern] Constitution': the *social* house is the place of humans, who can communicate via speech but 'find themselves in chains', ignorant of the world and its real mechanisms (ibid., 13). The house of *nature*, by contrast, is inhabited by 'nonhumans' that are governed by objective laws but unable to communicate (ibid., 13–14). Because the truth of reality is solely located in the second house, it is regarded as having the authority to 'tell the truth without being challenged, put an end to the interminable arguments [of the first house] through an incontestable form of authority that would stem from things themselves' (ibid., 14). This truth is communicated to the social house by 'travellers', who 'can make the mute world speak' (ibid.). Importantly, these travellers – the scientists – are the only ones who 'can go back and forth from one world to the other no matter what'; the only ones who have been able to escape the 'tyranny of the social dimension, public life, politics, subjective feelings, popular agitation' (ibid., 10).

Latour argues that the only solution to this problem is to get rid of the houses altogether, and to develop a new Constitution that allows for establishing *one* collective. Importantly, for Latour, constructivist critical approaches to reality (including, I argue, much of the Foucauldian discourse-focused approaches that I have previously critiqued) aim to cut off any ties between the natural and the social, closing the door for *any* travellers instead of throwing it wide open for everyone to go through. Constructivist approaches rigidify the binary between the natural and the social by leaving 'nature' to the scientists alone and by aiming for a rule of the social that is solely determined by social mechanisms and discourses, understood only by those who engage these mechanisms and discourses in theory and practice. In Latour's words (ibid., 33), the idea that "nature does not exist", since it is a matter of "social construction", only reinforces the division between the Cave and Heaven of Ideas by superimposing this division onto the one that distinguishes the human sciences from the natural sciences.

As Connolly (2017: 10) puts it in a related critique of what he calls 'sociocentrism', it is falsely assumed that 'cultural interpretation and social explanation can proceed without consulting deeply nonhuman, planetary forces with degrees of autonomy of their own.'

Instead, Latour aims to bring down the existence of the houses altogether, by giving external (natural) reality a (new) place in the work of the collective (ibid., 10). Rather than questioning the very existence of 'objective external reality', as the constructivists do, we should argue for the existence of *more* reality – a reality, however, that is never 'definitive'. Not because we do not yet know enough about it, but because it is ever-changing, depending on what human associates with what nonhuman at one particular moment in time. On the side of nature (as on the side of the social) we have 'countless imbroglios' that always already presuppose human participation (in the way nature is made sense of, for example, in laboratory experiments, in observation, in sampling etc.) (ibid., 20). For Latour (ibid., 89), the new collective should consequently be understood as 'an ever-growing list of associations between human and nonhuman actors'. This does not mean (and this is important, particularly regarding my previous critique of the New Materialist literature) that we should consequently strive for an indefinite openness towards embracing ever-new actors and human-nonhuman-associations. What we have to be open for, Latour maintains, are the *claims* for inclusion that are made by new associations that 'find themselves mobilized, recruited, socialized, domesticated' (ibid., 38), claims that the collective has to deliberate, but might well decide to turn down.

What the collective needs is an 'explicit procedure' in which decisions about collectivity, unity and in- or exclusion can be made together (ibid., 41). What it cannot do is simply 'adding together . . . nature and society' (ibid., 57). Everything needs to be thrown into the air and reassembled anew – which is a very slow process. Indeed, one of Latour's key messages in *Politics of Nature* is the need to slow things down, and to properly recognise everyone's right to fight for her/his/its own conception of 'the good common world': 'Nothing and no one must come in to simplify, shorten, limit, or reduce the scope of this debate in advance' (ibid., 130) – neither by asserting that the facts have been (scientifically) established, nor by asserting that nonhumans (and their scientific spokespersons) have to be excluded altogether.

However, despite creating space for the provision of 'right' over 'might' in a concrete political procedure, Latour is still unable to solve the problem of coloniality. Indeed, although Latour himself remarks at one point that the two-house logic is a specifically modern one in which '[n]on-Western cultures' have never been interested (ibid., 43), this argument remains at the level of statement: it is not accompanied by any analysis of how and why this logic had become dominant (in other words, it is not accompanied by an analysis, or even an acknowledgement of the need for an analysis of colonial exploitation). The politics of the collective remains an ideal scenario that is detached from any actual history of injustice and exploitation, or any concrete configurations of power. Falling once again prey to binary thinking, Latour opposes the *new* to the *old*, the *ideal* to grim *reality*. In the ideal scenario, decisions about inclusion or exclusion of particular propositions are justified if they follow 'due process' (ibid., 91). Does this mean, in practice, that propositions about, say, the need for reparations for past colonial exploitation can be excluded from public life if the deliberation of that proposition has been sufficiently 'slow'? Latour (ibid., 117) emphasises that the collective needs to recognise everyone's right to fight for the sort of collective world and life s/he/it desires, and that controversies should not be solved too quickly. At the end of the process of deliberation, the collective divides the world into 'friends' and 'enemies', and without that element of 'closure' it would not 'be able to learn' (ibid., 146). But all learning, reflection, judgement and decision-making remains internal to the collective, and though particular propositions themselves can point to questions of history, power and inequality, in theory the collective is still free to disregard history, exclude claims related to it and label them as 'enemies'. Of course the inability to judge from the outside is exactly what Latour thinks is desirable, and it is in line with my previous critique of those strands of critical scholarship that judge actually-existing political struggles against given theoretical schemata. However, as Harman (2014) well puts it, Latour's commitment to this sort of immanence 'verges on a commitment to victory' that is not interested in seeking justice for the 'losers of history' (ibid., 13–14). This is intolerable from a decolonial position, as it fails to recognise how modernity could only come about through being co-constituted with coloniality.[7]

In order to tackle this problem I want to argue that Latour's perspective needs to be complemented with a decolonial one, which, in contrast to the former, takes the existence of historical, differentiated structures of injustice and oppression as a starting-point for the critique of modern (binary) thought and for the development of alternatives. In other words, from a decolonial perspective questions of ontology are grounded in a (primary) analysis of history, though, as the next section will show, the latter inevitably points at the need for basing strategies of transformation in alternative 'outside' ontologies.

Starting from historical oppression: the problem of colonial difference

For decolonial scholars the colonial project has to be understood as co-constitutive of the project of modernity as such. Drawing on the philosophy of Enrique Dussel, Quijano introduced the term 'coloniality of power' in his 1989 foundational text (which was translated and published in English in 2007) in contrast to the term 'colonialism' that is meant to describe 'an explicit [historical] political order' (Quijano, 2007: 170). The concept of coloniality signifies the 'extension of Western capitalism' and the parallel 'extension of Western epistemology' to the non-European world, having started in the fifteenth/sixteenth centuries with the conquering of the Americas and 'the emergence of the Atlantic commercial circuit' (Mignolo, 2002: 58–9). Importantly, it is continuing to the present day as 'the most general form of domination in the world' (Quijano, 2007: 170). Because of the process of co-constitution, the beginning of modernity 'proper' likewise needs to be located in the fifteenth/sixteenth centuries and not, as it is often done, in the eighteenth century. Otherwise 'coloniality' becomes merely 'derivative'; serving to reaffirm modernity's claim to be all-inclusive and universal, erasing or relegating the 'Iberian foundational period of capitalist expansion and coloniality . . . to the Middle Ages as the Black Legend' (Mignolo, 2002: 61).

Because of the crucial epistemological dimension of the modernity/coloniality project, decolonisation needs to include decolonising knowledge, which for Mignolo (2007) needs to happen through an epistemic 'de-linking' of the Global South.[8] For this argument Mignolo draws on Latin American dependency theory, which he defines as 'a political statement for . . . social transformation *of and from Third World countries*' (in contrast to world-system analysis which Mignolo uses as well, but which he defines as 'a political statement for academic transformation *from First World countries*') (Mignolo, 2002: 63; emphasis added). The decolonising task lies in a move towards 'the re-construction and restitution of silenced histories, repressed subjectivities, subalternized knowledges and languages performed by the Totality depicted under the names of modernity and rationality.' (Mignolo, 2007: 451)

In short, for decolonial scholars we need different, non-Western ('de-linked') epistemologies that are able to go beyond modern (binary) categorisation because of the actual, historical (colonial) forms of oppression that have brought about the latter. This is what makes their argument relevant for the purposes of this book. In order to move beyond modern/colonial binaries in environmental activism, it is not enough to think beyond modernity *from within the inside of European modernity itself*. In other words, it is not enough to engage in 'intra-modern' (Escobar, 2007: 180) or 'Eurocentric' critique (Dussel, cited in Mignolo, 2002: 57). Instead, new thought has to emerge from the *other* side of what Mignolo calls the 'colonial difference', which is the 'difference between center and periphery, between the Eurocentric critique of Eurocentrism and knowledge production by those who participated in building the modern/colonial world and those who have been left out of the discussion' (ibid., 63). To paraphrase Gurminder K. Bhambra (2014: 130), rather than an endogenous understanding of the development of European modernity, Europe should have an exogenous one that acknowledges how it has evolved from the oppression of 'other cultures [that] constitute the ground of European self-realization'.

Besides decolonising knowledge, we also need to decolonise Being – which brings us back to the realm of *ontology*. The term 'coloniality of Being' was initially used by Mignolo and then taken up and further developed, as a concept, by Nelson Maldonado-Torres (2007). While 'coloniality of power' refers to 'the interrelation among modern forms of exploitation and domination (power)', and 'coloniality of knowledge' to colonial modes of knowledge production as

previously outlined, 'coloniality of being' is about 'the lived experience of coloniality' of those who have been subjected to it (ibid., 242). Drawing among others
on Emmanuel Lévinas, Maldonado-Torres critically interrogates the exclusionary
dimension of ontological thinking as a 'philosophy of power' when it is taken 'as
foundation or ultimate end'. Critiquing Heidegger, Lévinas argues that the problem of a focus on ontology is that it 'gives priority to an anonymous Being over
and beyond the [ethical] self-Other relation'. Consequently, as seen in Heidegger,
the yearning becomes one for 'authenticity' rather than 'radical responsibility'
(cited in ibid., 258).

Due to understanding being (matter) as always exceeding any capture, and as
disrupting any sense of coherence or identity, which, it could be argued, leads to
an *extension* of responsibility towards the Other rather than a neglect, this critique does not fully apply to New Materialists. As Mario Blaser (2013: 551) has
pointed out, the 'foundational' (and thereby exclusionary) claim of ontology can
be avoided if we go beyond *both* a notion of ontology as a mere heuristic device,
and as the making of a foundational claim. Rather, Blaser (ibid.) argues, we should
regard it as a 'heuristic device' that contributes to 'enact[ing] the "fact", which
allows us 'to articulate a foundationless foundational claim.' In other words, we
do stick to the idea of there being a 'reality out there', but, similarly to the New
Materialists (on whom Blaser indeed draws as well; ibid.), we do not regard this
'reality' as ever graspable with one definite set of concepts or categories. Ontology then becomes 'a way of worlding, a form of enacting' a reality that nevertheless, in some sense, *is* – but as 'always in the making through the dynamic
relations of hybrid assemblages' (ibid., 551–2). Because this involves, for Blaser,
'a certain political sensibility' that contains some normative commitment to enact
reality as a 'pluriverse', Blaser (ibid., 552) calls this *political ontology* (see Chapter 2 for more details).

However, taken on its own, this still does not get us away from the question
whether *some* 'enactments' of ontology should be prioritised over *others*. In other
words, before we move to the *pluriverse*, we need to come to terms with the colonial difference that has led to the dominance of the (modern) *universe* in the first
place. Coming back to Maldonado-Torres's Lévinasian question about 'radical
responsibility', what the New Materialist approach does not offer is the ability
to reflect on *concrete* responsibility towards *concrete* others, resulting from the
analysis of a *concrete* history of oppression and annihilation. This is precisely
why we need the decolonial approach: it asks for what Mignolo (2002: 63) calls
a 're-construction and restitution' of the knowledge and, indeed, 'being' (as lived
experience) of those who have been silenced. This does not mean that there is
something ontologically 'original' that needs to be restored (like an original pre-
colonial ontology) (Escobar, 2007: 186). Instead, decoloniality functions as 'an
invitation to think modernity/coloniality critically from different epistemic positions and according to the manifold experiences of subjects who suffer different
dimensions of the coloniality of Being.' (Maldonado-Torres, 2007: 261)

Mignolo (2007: 452) is keen to distinguish the *de-* from the *postcolonial* critique by emphasising the more fundamental 'fracture' that the former introduces

into a Eurocentred modernity. For Mignolo (ibid.), postcolonial theorists remain too wedded to European thought (particularly European postmodern perspectives) and are insufficiently radical when it comes to the need for transformation beyond the academy and the realm of culture. However, though there has so far been little attempt to bring them together, both perspectives are not mutually exclusive (Bhambra, 2014: 119). As I will show in this book (particularly in Chapter 3) a postcolonial perspective is able to tackle the latent assumption in some decolonial work that social movements from the Global South think and act from the 'outside' qua definition. This is related to the problem that foundational decolonial texts often base their core arguments in a rather sweeping, 'easy' historical and geopolitical account in which a clear (geographical) demarcation is made between the 'inside' and 'outside' of modernity, at least when it comes to practical examples (decolonial scholars are more successful in avoiding this problem in their conceptual work). As Julia Suárez-Krabbe (2016: 17–18) has pointed out, a related problem is that decolonial scholars take their own (scholarly and activist) work as coming from a position of subalternity simply by the fact that they are located or have roots in the 'South', without further reflection on how Latin American 'elites' (including scholarly elites) have been constituted as part of the continent's own, complex colonial trajectory. We might need the 'pessimism' of postcolonial scholars, who are often more critical in relation to what is going on in social movements in the Global South itself, in order to counter the sometimes overly euphoric, liberation-affirming 'optimism' of the decolonial critique.[9]

In line with a more postcolonial approach, this book continues to engage the concepts and epistemologies of European thinkers: in particular, as already mentioned, the thought of Bruno Latour, but also the philosophy of Gilles Deleuze. Like Latour, Deleuze is guilty of not linking the problems of modernity as well as his (alternative) metaphysics of non-binary transformation explicitly to the project of coloniality, and of once again pursuing the development of philosophical alternatives utterly unrelated to actually-existing 'other' philosophies. In order to avoid the reproduction of the coloniality of knowledge as much as possible while still drawing on these two thinkers, I will only consider Latour and Deleuze to the extent that their ontologies, epistemologies, concepts and methodologies are supporting thought and practice from the 'outside' of modernity, which will be established by bringing them into dialogue with the de- and postcolonial thought of Sylvia Wynter, Gayatri C. Spivak and María Lugones, as well as anti-GMO argument and activist at both European and non-European sites. Out of all this will emerge different, decolonised ways of understanding nature, nonhumans, humans, the sciences, 'voices' and (global) political activism – moving through and then beyond modern/colonial binaries.

Chapter outline

In Chapter 2 I will analyse the arguments of anti-GMO advocates at the site of science itself. I will show how the activists' drawing on complexity science in order to reject the technology of GE has the potential of challenging modern/colonial binaries more generally, by questioning the central modern distinctions between subject and object, mind and matter, human and nonhuman. However, I will also argue that relying on science does not take us far enough in our attempt to over-come modernity/coloniality, because even those advocating alternative scientific approaches continue to regard science as the privileged site of providing informa-tion about the objective 'truth' of nature. This leaves the crucial nature/society dichotomy intact. In order to address this problem, I will transform some of the claims made by anti-GMO activists, as well as the claims made by their opponents, into what Latour (2004) calls *propositions*. Propositions, contra to facts, arise out of particular human-nonhuman-associations. They make a claim for inclusion into the *collective*, with the latter deliberating these claims in 'due process' (ibid., 91) without short-cutting discussions by referring to facts. Analysing the claims of five propositions – 'survival machine' (proposition 1), 'sea abyss descendants' (proposition 2), 'emergent relationality' (proposition 3), 'controllable possession' (proposition 4) and 'dancing whole' (proposition 5), I will outline the 'ontological conflict' (Blaser, 2013: 548) that is taking place among them: some propositions affirm modern/colonial binaries, others are able to move away from them at least to a certain extent. Drawing on Blaser's concept of political ontology, I will show that using the procedure of Latour's collective in this way is already in itself a means of contesting the universalist claims of modern/colonial (binary) ontol-ogy, by 'shrinking' the latter and making it just one particular ontological 'enact-ment' among others (ibid., 553). This particular 'method' puts my chapter in line with the general thrust of decolonial critique and ethos. Moreover, concerns over coloniality enter the 'story' (ibid., 552) of the collective in this chapter through the introduction of various 'witnesses' and 'spokespersons' that speak on behalf, or against, various propositions; shedding light, for example, on the coloniality of proposition 1 'survival machine' (which is based on Richard Dawkins's claim about the gene as 'selfish' and the organism as a 'survival machine'; cf. Chapter 2). All of this will lead me in the end to affirm, from a decolonial position, those propositions that more fundamentally question modern/colonial binaries, such as proposition 5 'dancing whole' (which depicts the organism as a self-coordinated dance that cannot be fully understood, let alone controlled). However, further drawing on Latour's concept of the spokesperson, I will also argue that in order to fully decolonise science-based anti-GMO arguments, activists need to give up the idea of being able to 'truthfully' represent nature as one harmonious entity. Instead, as I will show based on a concrete example, they need to start measuring the 'faithfulness' of their own testimony by looking at the extent to which nonhu-mans are able to resist and transform the questions activists ask of them.

Chapter 3 will engage with the anti-GMO argument that agricultural biotech-nology is yet another means used by MNCs, international organisations and West-ern states to push through their neocolonial and neoliberal agenda. Taking the controversy around Bt cotton in India as example, I will analyse the binary logic that underlies the argument of prominent environmentalist Vandana Shiva, who distinguishes between the modern, neoliberal, neocolonial way of doing agricul-ture on the one hand, and a traditional, authentic, Indigenous way on the other.

The same logic is found, as I will show, in the argument of Shiva's principal opponent, Ronald J. Herring, who understands Bt-adopting farmers as rational agents making decisions about what to plant on the basis of 'what works'. I will argue that both sets of argument neglect to what extent the actual 'voices' of farmers are shaped by particular historical-colonial conditions – not just socioeconomically, but also at a deeper, onto-epistemological level. Drawing on Spivak's famous essay 'Can the Subaltern Speak?' (1988) and bringing the latter's take on representation into conversation with Latour's understanding of spokespersonship, I will develop a new understanding of how to give an adequate 'voice' to both subaltern humans *and* nonhumans that are involved in the Bt cotton controversy. This will be fleshed out further with the help of Deleuze, particularly his concept of the 'statement' (Deleuze, 1999) and of the notion of 'regimes of signs' that Deleuze developed together with Guattari (Deleuze and Guattari, 2004). Deleuze, I argue, enables us to move towards an alternative understanding of reality that can only be accessed through *sense*, rather than through a narrow focus on and rational processing of what is actually 'said' or 'done'. Based on the example of the Bt controversy, I will argue that we need to develop new 'statements' about GMOs and environmental protest. These 'statements' emerge, on the one hand, out of having to endure the *silence* that we find in relation to alternative onto- epistemologies, a silence that is related to India's particular postcolonial history that, in relation to the role of science, progress and political economy, has emerged out of *continuity* with its colonial history. On the other hand, new 'statements' also arise out of the 'noise' of Indian anti-GMO activism. As I will show, GMOs have become part of particular historical-social-natural-economic assemblages that have rendered unfamiliar those objects that neoliberal forces of globalisation have previously (successfully) rendered too familiar to contest. However, I will also critique the ongoing activist fixation on the ontological properties of GMOs as intrinsically 'dangerous', the continuing strong reliance on scientific 'facts', and the affirmation of an (imagined) traditional, familiar, authentic identity that risks cutting down the human-nonhuman-assemblage to a form that once again correlates with what I will call the modern/colonial apparatus of Man/gene/State.

Chapter 4 is interested in the connections that anti-GMO activists from various political, geographical, socioeconomic and cultural backgrounds make with each other in order to transform their protest into a 'global' one. In order to achieve this activists tend to focus on the shared desire for global justice, the rejection of neoliberalism and the opposition to the global strategies of agrobusiness. The academy often understands the spatiality of 'global' resistance as being either one of networked connection, or of protest taking place within specific, locally bounded spaces that are purposefully designed as 'living the change' for which activists fight (see for example the Occupy protests). I will use Lugones's (2003) work in order to disrupt and decolonise these spatial imaginaries, drawing on her understanding of resistance as having to take place from the site of the multiple, 'impure', 'world'-travelling self against oppressions that need to be understood as enmeshed. Based on that approach, I will critique the coloniality of one particular historical example of global protest: the 1999 Intercontinental Caravan (ICC) with which several hundred Indian farmers and other non-Western activists travelled around Europe together with European activists in order to protest at the sites of international organisations and corporations. The coloniality of this protest manifests itself, I argue, in the way that activists seek harmony and 'sameness' among themselves with regard to their identity, objectives and overall strategies, as well as (relatedly) the particular expectations that European activists had in relation to the 'true' peasant identity of non-Western participants. Using Lugones's concepts of the pilgrimage and the streetwalker, the chapter will then develop alternative, decolonial understandings (and spatial imaginaries) of the potential for resistance of the ICC (and global protest movements more generally). I will argue that giving greater space to nonhuman realities enables us to make better sense of oppressed experience, and to subordinate our thinking about objectives and strategies to the understandings that emerge out of long-winded, intersubjective, (human and non-human) body-to-body encounters. The latter, I maintain with Lugones, have the potential of challenging our (colonial) frames of understanding domination, our own identity and, most importantly, our perception of the Other.

Finally, given that this book has emerged out of a project in transition (cf. the Preface), the Conclusion will serve to provide a retrospective reflection on the themes that have come to the fore while writing the book. It will make new connections between those themes, both in relation to anti-GMO argument and practice, and in relation to the different conceptual resources that I have used throughout. Because this book also serves to make practical suggestions about how to concretely move forward in environmental/anti-GMO activism, the Conclusion will end with a seven-point activist 'manifesto'.

Notes

1 That said, a 2014 YouGov poll showed that 41 percent of the British population remain sceptical about GMOs, with only 17 percent being openly in favour (Jordan, 2014). Similarly, according to UK Government polling in 2015, only one in three of the British public supported nuclear power at that point, which was lower than in previous years (Department of Energy & Climate Change, 2015). However, what interests me is the particular reasoning that environmentalists such as Monbiot and Lynas provide, not to what extent this belief is anchored in the beliefs of the general public.

2 I use the term 'non-modern' as it is employed by María Lugones (2010). Drawing on Juan R. Aparicio and Mario Blaser, Lugones defines 'modern' as denying 'the challenge of the existence of other worlds with different ontological presuppositions' (ibid.: 749). The 'non-modern' is precisely what is denied and yet becomes co-constitutive. Lugones uses that term in contrast to the concept of the 'pre-modern', which, she says, is already subject to the modern, categorising logic (ibid.: 743).

3 It needs to be acknowledged though that Chandler and Reid's (forthcoming) article rightly critiques the problem of focusing *solely* on Indigenous knowledge as a question of ontology. Chandler and Reid very well point out how such an approach is in danger of ignoring the actual state of dispossession, dependency and brokenness of Indigenous communities, caused by decades of exploitation. However, decolonial scholars are well aware of that – they do not confine their analysis of power to the cultural realm but precisely argue that the coloniality of power is about the intertwining of epistemological and socioeconomic oppression in the project of modernity (see e.g. Quijano, 2007).

4 The generality and metaphysical dimension of this challenge is precisely what other scholars interested in new understandings of matter/materiality and nonhuman agency critique. See for example Bruce Braun and Sarah Whatmore's (2010) questioning of what they call Jane Bennett's metaphysical vitalism, which they contrast with the suggested need for an analysis of the 'specificity of the matter at hand' (ibid.: xxix–xxx). See also Abrahamsson, Bertoni and Mol (2015) for a similar critique. However, though this is a worthwhile and interesting controversy, my aim of challenging the coloniality of knowledge makes me turn to precisely those scholars who are interested in wider philosophical development rather than ethnographic specificity.

5 It needs to be said that in *Facing the Planetary: Entangled Humanism and the Politics of Swarming* Connolly (2017) engages in great detail with forces of domination, and also draws in an exemplary manner on non-Western thought and perspectives for that purpose. His book overall provides an impressive, successful attempt to tackle the problem of human exceptionalism and what he calls 'sociocentrism' that is found in both those approaches that see planetary forces as an 'environment' for human existence as well as those environmentalists that disregard the autonomy of nonhuman forces beyond human impact. However, various problems remain. One of them is Connolly's occasional tendency to reify and take as authoritative the truth-claims of scientific 'discoveries' and 'new' scientific knowledge. Secondly, despite Connolly's attempt to reach out to non-Western scholars and knowledges, it is striking that this is largely confined to the last two chapters in which he fleshes out the practical implications of his politics of swarming. Indeed, the conceptual work in the first four chapters remains confined to an engagement with largely Western scholars and Western scientists – despite the overlaps that much of Connolly's writing here has with Indigenous thinkers such as E. Richard Atleo (Umeek), Vine Deloria Jr, Taiaiake Alfred, or Paula Gunn Allen. Zoe Todd's (2016) critique of Latour as failing to engage any Indigenous conceptions of 'Gaia' applies, in that sense, also to Connolly's concept of 'entangled humanism'.

6 Some scholars might object that the concept of 'spokespersonship' once again focuses too much on human agency.

7 This also raises a further question about how we might understand the relation between Latour's constitutionalism and the significance of the constitution in French political history (cf. Barry 2005).

8 'De-linking' replaces Mignolo's earlier emphasis on the need for 'border thinking'. For the latter see Mignolo (2000).

9 I am grateful to Rahul Rao for putting it that distinctly in a conference panel discussion.

2 'No one knows what an environment can do'

From facts to concerns in the GMO controversy

> The opponents of GMOs would like us to believe that genetic modification of crops has dangerous implications for our food supply. But the reality is that there is no evidence that it is inherently any more dangerous or riskier than any other plant breeding technique. It is merely another application of science to agriculture that is designed to help feed a growing world population.
>
> Steven Cerier (2016)

> Remarkable people like Vandana Shiva and Chee Yokeling of the Third World Network . . . taught me just how important science is in shaping people's lives and how crucial to get the science right. To me, science is the most intimate knowledge of nature that is beautiful beyond compare; it is also reliable knowledge that enables us to live sustainably with nature, and I have dedicated my life since to defending and promoting that science.
>
> Mae-Wan Ho (2013)

Above statements were made by two people who hold diametrically opposed views on the question of agricultural biotechnology. While Steven Cerier emphatically calls for the public to recognise that GMOs are as safe to produce and consume as any other agricultural product, the late Mae-Wan Ho was a devoted anti-GMO advocate and activist. But as the quotes illustrate, what unites both is their belief in the role that science should play in the shaping of society and the latter's engagement with its natural environment: both agree that science is able to truthfully understand and represent nature, and that scientific evidence should be the decisive benchmark criterion for making decisions about how to live and act. What they disagree on is what kind of science generates such evidence. Science can be be done in the *right* or the *wrong* way, resulting in *facts* or *false claims*. As I have indicated in Chapter 1, this understanding of the role of science is inextricably linked to modern/colonial thought, and the distinction that is made between nature and culture.

How can we properly move beyond this understanding? Can there be a different, decolonised, non-modern role for science? A different understanding of truth? Can and should nonhumans have a 'voice' in the GMO controversy, and if yes, in what way? And do in fact all *humans* have a voice?

As I have already mentioned in Chapter 1, for Latour we can only move beyond what he calls the old (modern, binary) Constitution once we are prepared to give up the notion of 'facts' without simultaneously giving up on the existence of external reality. Rather than questioning the very existence of 'objective external reality', we should argue for the existence of *more* reality – a reality, however, that is never definitive. Not because we do not yet know enough about it, but because it is ever-changing, depending on what human associates with what nonhuman at what particular moment in time. As Latour explains in quite some detail in *Politics of Nature* (2004), acknowledging reality as ever-changing does not mean that whatever association comes to the fore needs to become part of what he calls the collective (cf. Chapter 1). Indeed, the collective has to decide in 'due process' (ibid., 91) whether to include or exclude particular propositions. Propositions are 'new and unforeseen association[s]' that bring with them 'uncertainty and not arrogance': 'a river, a troop of elephants, a climate, El Niño, a mayor, a town, a park, have to be taken as propositions to the collective' (ibid., 83). Propositions emerge from human intermingling with an external reality that is not simply 'there', that is not simply 'fact', but that comes on us as 'surprises and events': humans and nonhumans 'emerge in surprising fashion, lengthening the list of beings that must be taken into account' (ibid., 79). In line with the general thrust of Latour's work, propositions are not characterised by an internal substance but by the force of their relations, and they do exist as actors only by having an effect (cf. Harman, 2014: 41). The crucial modern binary of *subject* (as that which makes a statement about someone/something else) and *object* (as that about whom/which a statement is made) is done away with (Latour, 2004: 83). Science, for Latour, is an important means of constantly bringing new propositions to the attention of the collective, due to the intermingling of humans and nonhumans that takes place in scientific experiments and investigations. However, the role of science should be precisely confined to this: bringing to our attention and making a case for the need to include into the collective new propositions – a case that then needs to be deliberated by the collective in relation to other propositions (ibid., 111). New propositions make a claim for inclusion, they are not 'facts' that can close down discussion. As this chapter (and the book more generally) will show, in the GMO controversy a majority of advocates on both sides does not truly give up on facts: most biotech proponents and opponents continue to believe that nature is something 'out there' that can be made sense of by appropriate scientific investigation. Both believe in nature as teaching us *facts* of existence via science.

However, as already pointed out in Chapter 1, Latour's depiction of the collective as an abstract ideal in which the right 'procedure' and the setting up of the right institutions both trumps an analysis of power and domination and an engagement with the actual *substance* of the claims for inclusion that are made is deeply problematic. Latour assumes that all claims for inclusion come from the same equal-level starting-position, that they have no history and that they all have to be considered in the same way. Pointing at a history of domination and exploitation can potentially be used by a particular proposition to back up its claim for inclusion, but it is still possible to justly exclude the respective proposition if 'due process' is adhered to. In contrast to his earlier work, which was all about 'might' winning over 'right' (in the sense of actors either succeeding or failing in assembling and connecting to allies; cf. Harman, 2014: 13), Latour's collective is indeed about what is 'right' – but at a merely procedural level. This is obviously in line with Latour's more general attempt to eradicate any possibility of appealing to a Truth that lies outside of what happens in the relation, connection and struggle of multiple actors affecting each other. Although that aversion to Truth makes Latour critical of thought that is not only modern but also colonial, it also makes it impossible to treat coloniality and colonial violation as anything other than a claim that needs to be deliberated and negotiated according to a procedure of the collective that is assumed to sit outside of that very violation.

In this chapter I argue that it is possible to address this problem without having to give up on the merits of Latour's critique – and his politics of the collective – altogether. I will do so by understanding the collective and its procedure as an enactment of what Mario Blaser (2013) calls *political ontology* (cf. Chapter 1). As Graham Harman (2014: 56) rightly points out, despite Latour's focus on 'right' procedure and institutions in *Politics of Nature*, Latour still treats politics (at least implicitly) as ontological. In other words, the procedure of the collective is not just about making decisions about one specific realm – the political – but it tells us something more generally about the 'being' of the world. One of the most striking examples for this tendency can be found at the end of the last chapter of *Politics of Nature*, where Latour (2004: 220) says that 'the diverse' can only become visible 'if it no longer sets itself apart against the prematurely unified background of nature'. In that case, Latour says, '[r]elativism would disappear with absolutism' and what would remain were to be 'relationism, the common world to be built.'[1] Instead of finding this ontologisation problematic, as Harman does, I argue that it can indeed be useful for the task of *decolonising* the politics of the collective, and, going along with that, the politics and argument of science-based anti-GMO activists such as Ho.

For Blaser (2013: 551) we need to move away from the standard understanding of ontology as defining a 'being' (in the singular) that only allows for multiplicity in the perspectives that are possible on that 'being'. Likewise we should not use the concept of ontology as a mere 'heuristic device' that points to the limits of our own understanding of the world, as it is often done in anthropology today (ibid.). In practice, the latter mostly leads to the implicit claim that no reality is fully 'real', since all ontologies are merely relative to each other – a move that empties the concept of ontology of all substance. Blaser (ibid., 551) argues that instead we need to understand ontology as making a 'foundationless foundational claim'. Blaser embraces an understanding of reality that can only become 'real' in and through 'enactments' or 'worldings', of which there are always multiple ones (ibid., 552). In other words, referring to multiple ontologies does not mean that there are multiple perspectives on reality, or alternatively multiple realities that are not really 'real', but that in their enactment *all* (multiple) realities are fully real (ibid., 551). In ways that are closely related to Latour (on whom he indeed draws, among others), Blaser (ibid., 551) maintains that this is only possible if we arrive at a 'material-semiotic formulation' of reality that bypasses the nature/culture divide altogether by arguing that 'reality is always in the making through the dynamic relations of hybrid assemblages', and by insisting that it can never be accurately represented by the mind (ibid., 551–2).

However, in contrast to Latour, Blaser explicitly *politicises* this understanding of ontology by grounding it in a decolonial, historical analysis of the ways in which the universalist claim of modern ontology has evolved from the encounter and suppression of other ontologies. Political ontology demands a 'political sensibility' that leads to a 'commitment to the pluriverse' (ibid., 552). This commitment actively seeks a 'shrinking' of modernity by 'mak[ing] it something more specific and contrastable, thus liberating the conceptual-ontological space for something else to exist.' (ibid., 553). By arguing that Latour's collective is precisely grounded in political ontology and the commitment that goes along with it, it becomes possible to link the existence and procedure of the collective to the critique, ethos and political objectives of decolonial critique. By regarding *all* propositions as being of equal worth, and as having the right to making a claim for inclusion, the collective becomes able to disrupt modern/colonial thought as such, precisely because the claims of the latter only become *specific*, 'shrunk' claims that compete with others at an equal level. The procedure of the collective becomes a 'story' that puts political ontology into practice by participating in the 'making' of reality; 'open[ing] up a space for, and enact[ing], the pluriverse' (ibid., 552–3).

Nevertheless, decolonising the politics of the collective still requires a further step, namely an acknowledgement of the fact that modern/colonial and 'other' claims have emerged from a history in which the latter have been oppressed and excluded by the former. This becomes possible by not only regarding the collective itself as an enactment of political ontology, but by 'making' it also into a site on which 'ontological conflicts' are carried out. As Blaser (ibid., 548) argues, ontological conflicts 'involve conflicting stories about "what is there" and how they constitute realities in power-charged fields.' Depicting the conflict between modern/colonial ontology and 'other' ontologies[2] as being carried out through the procedure of the collective allows me to not just enact a reality in which modern ontological claims are just one set of claims among others. It also allows me to tell the story of coloniality, by pointing out how modern claims have served to oppress what is 'other', and to become all-encompassing in the process of doing so.

It is the aim of this chapter to tell a 'story' about the scientific dimension of the GMO controversy and its modern/colonial ground through 'enacting' the procedure of the collective. More concretely, the story will be about the claims for inclusion into the collective made by five propositions: 'survival machine' (proposition 1), 'sea abyss descendants' (proposition 2), 'emergent relationality' (proposition 3), 'controllable possession' (proposition 4) and 'dancing whole' (proposition 5). All of these propositions are linked to the GMO controversy, emerging from the arguments and standpoints of both those who promote as well as those who oppose it from a scientific point of view. Some propositions are linked to what Latour calls the old Constitution (i.e. modern/colonial thought), others attempt to question the latter and free up space for 'other' ways to think and be (though with varying degrees of success). Each claim and proposition will be accompanied by the testimonies of what Latour calls 'witnesses' or 'spokespersons', who speak either in favour of or against the propositions. According to Latour (2004: 24), spokespersons are not 'intermediaries' between two sides, but 'mediators'. Mediators 'transform, translate, distort and modify the meaning or the elements they are supposed to carry' (Latour, 2005: 39). Understanding those who speak as 'spokespersons' troubles both the claim that it is possible to fully and accurately represent someone or something else and the idea that representation is nothing but a construction and idealisation (ibid; Latour, 2004: 70). The testimonies, as acts of mediation and translation, aid the collective in making a decision about which propositions to in- and which to exclude. Instead of a conclusion, the chapter will end with a reflection on what the 'new' collective looks like after this process of deliberation, and what the concrete consequences could be for an anti-GMO activism that is able to move beyond modern/colonial binaries.

Man/gene

Proposition 1: 'Survival machine'

> We are survival machines – robot vehicles blindly programmed to preserve the selfish molecules known as genes . . . another survival machine . . . is something that gets in the way, or something that can be exploited.
>
> Richard Dawkins, *The Selfish Gene* (1989): vii, 66

In the nineteenth century Gregor Mendel conducted experiments with several generations of plants that led him to claim that each organism has specific 'units' that pass on its traits to the next generation (Watson, 2004). In the 1940s and 1950s, Rosalind Franklin and, after her, James D. Watson and Francis Crick located these units in the organism's deoxyribonucleic acid (DNA) (ibid.). Today a gene is commonly defined as

> a sequence of DNA in a defined location of the genome . . . that specifies the amino acid sequence of a polypeptide via non-overlapping triplet code. . . . The coding sequence is flanked by regulatory signals for transcription to start at one end and stop at the other.
>
> (Ho, 2010: 84)

In his book *The Selfish Gene* (1989), Richard Dawkins gave a famous 'speech prosthesis' (Latour, 2004: 67) – the one of the 'survival machine' – to the gene and the organism that the latter is supposed to build. Proposition 1 'survival machine' pursues the preservation and passing on of its genetic make-up as its primary aim. The struggle for survival takes place in a hostile environment in which the 'machine' has to compete with others for scarce resources, resulting in exploitation. Proposition 1 'survival machine' is in line with mainstream understandings of genetics: the gene sits at the centre of the association and passes on information in a linear manner to the protein that then makes (together with other proteins) the organism. The so-called 'Central Dogma' of traditional genetics relies on a Newtonian view of the world as existing in perfect equilibrium, with linear cause-and-effect relations. Proposition 1 'survival machine' consequently draws on what Latour calls the old Constitution: the house of nature is understood to be *one*, governed by *one* set of rules that is as valid for the organic world as it is for the inorganic one researched by physicists.

The first witness

For critical scientist Susan Oyama (2000: 1), a central feature of traditional genetics is its reliance on making an ontological distinction between the 'being' of information and the process of its realisation. In making this distinction, one part of the organism (the information as 'being', in the genes) is cut off from the rest and allocated the function of eternal, atemporal developmental 'cause' (ibid.). In doing so, the question of life continues to be asked with reference to its origin, no matter whether the latter is located in a causalistic God, or (nowadays) in the gene as 'Nature's agent' (ibid.). The underlying Western world view that emphasises 'the persistence of the eternal soul, or order and stability in the face of change' (Ho, 1998: 72) is contained in the notion of what August Weismann originally called the 'germ plasm', which assumes that biological heritage is contained and passed on unaltered through the generations. If that notion were given up, it would become difficult to explain how disorder (a 'heap of chemicals') could produce the ordered unity of being (Oyama, 2000: 14). 'Gene centrism' strongly links to the ideal of predictability and (self-) control. The stated aim is to get to know life as such and thereby oneself, with this knowledge being compressed so that it can be possessed and controlled by a (universal) subject. Evelyn Fox Keller (2000: 6) refers to Walter Gilbert, who describes as ideal the possibility that in future every man and woman is able to pull out a CD of his/her pocket that includes every sequence of his/her DNA. This ideal mirrors a reductionist account of life: what we can and should know about ourselves, indeed the whole 'grail' of knowledge, is supposed to be contained in the sequences of our basic units (ibid.).

Proposition 2: 'sea abyss descendants'

> [We are] the descendants . . . made [by an] experience [of] you, original victim floating towards the sea's abysses . . . the Africans who lived through the experiences of deportation to the Americas, confronting the unknown with neither preparation nor challenge . . . wrenched from their everyday, familiar land, away from protecting gods and a tutelary community.
>
> Éduoard Glissant, *Poetics of Relation* (1997): 5, 8–9

Unlike proposition 1 'survival machine', which defines its identity as fixed and eternally true, the above proposition has a temporal dimension. It locates its identity historically, calling upon the collective to recognise an existence that, as the next witness will show, has been historically denied. While proposition 1 'survival machine' grounds its existence in the old Constitution, proposition 2 'sea abyss descendants' grounds its claim for inclusion in the need to reject that Constitution, which, as the witness will argue, eradicates proposition 2's very right to ontological self-definition (cf. Latour, 2004: 179).

The second witness

As Sylvia Wynter's (2003) argument makes clear, proposition 1 'survival machine' follows the logic of what Walter Mignolo (2000) calls 'the foundational "colonial difference" on which the world of modernity was to institute itself' (Wynter, 2003: 260). As I have already outlined in Chapter 1, this means that modernity proceeded to understand its own development independent of coloniality. The way modernity in itself arose from the encounter and consequent oppression of difference was systematically denied and disguised by the former's all-encompassing, universal claims.

Oyama's point about the continuity of dualistic conceptions in traditional genetics (God/Nature/gene creating the world/the organism) relates to Wynter's point about the continuity of the modern 'master code' with its medieval predecessor. While the medieval theocratic code was based on an ontological 'Spirit/Flesh' divide between the Clergy and the lay people who remained bound to Original Sin (ibid., 263), Renaissance humanism replaced this with a 'Reason/Unreason' divide. Man became perceived as created by God with the capacity for reason, while 'irrational' animals and natives were unable to free themselves from the chains of nature and therefore unable to follow (God-given) Natural Law (ibid., 286). Wynter calls this the creation of 'Man1', while 'Man2' (from the eighteenth century onwards) deleted God altogether and replaced it with Darwin's laws of evolution and natural selection (ibid., 264). 'Nature', according to these laws, equipped the different races of 'Man' – now a purely biological being – differently, allowing white Man to explain his own flourishing by referring to his biological superiority, which became manifest in his capacity to adapt. The misery of other 'races' could consequently be explained by them not having adapted quite so well, rather than by their active exploitation through white Man (ibid., 323).

Wynter (as indeed all decolonial critics) puts a lot of emphasis on the historical context of this development. She argues that it can only be understood in relation to the colonial conquest and exploitation that started with the 'discovery' of the Americas in 1492. At the same time as Man was freeing himself from the shackles of the Church, He encountered its 'Human Other' (ibid., 291). But while the old Spirit/Flesh code had the potential of assembling, under the concept of Fallen Man, all kinds of different human beings – at least as long as those human beings were part of, or converted to, the True Religion – modern Man conflated Himself with the idea of humanity as such (ibid., 281). This led to what Wynter (ibid.) calls the 'overrepresentation' of Man, who could see other humans only 'adaptively, as the lack of the West's ontologically absolute self-description.' Unable to extricate Himself from medieval Christianity's 'monotheistic conception', Man could not conceive of a 'Human Other' as being different, but still human: Man is One, as God was/is One (ibid., 291). The resulting ontological absolutism defined the essence of the 'Human Other' as one of lack. This stands in contrast to what Latour (2004: 179) defines as decisive for the existence of the collective: even though the collective hierarchicises and excludes claims in order to function politically, it grants every association (included or excluded) the right to ontological (positive) self-definition.

Man's overrepresentation went hand in hand with nature becoming an 'autonomously functioning force', which in turn went hand in hand with sweeping away the medieval conception of the nonhomogeneity of Earth and the Heavens (Wynter, 2003: 264). While in medieval times Earth (as fallen) was ontologically distinct from the perfection of the Heavens, the emerging Newtonian view that proposition 1 'survival machine' signs up to argues that 'all parts of the universe [are] made of the same forces, of the same matter' (ibid., 281). Assumptions about homogeneity became a decisive organising principle for both Man and the study of physics. But what took place alongside these assumptions was 'large-scale expropriation and mass enslavements . . . on the grounds of a naturally determined difference of rational substance between [enslaved people] and their expropriators and slave masters' (Man1) (ibid., 304). This was then later replaced with a racialised divide that defined the slave masters as better equipped to survive and flourish in a world characterised by scarcity (Man2) (ibid., 266, 307). In other words, proposition 1 'survival machine' is, against its claims, far from being 'natural' or an atemporal 'fact': it is overrepresented Man; it is *Man/gene*, the identity claim of which is/was based on a history of exploitation and eradication of the 'Human Other' (cf. ibid., 265).

Stepping into the witness stand, Wynter is able to show how propositions that are put to the collective have to be looked at in a historical context. Her witness testimony enables us to pull together Latour's critique of the old Constitution and decolonial critique in a context of both historical and contemporary political contestation. Proposition 1 'survival machine' – Man/gene – is part of the old Constitution (or, as Blaser would call it, modern/universalist ontology), and it is that Constitution that grounds colonial exploitation and oppression, past and present. Wynter allows us to make a judgement about and between both propositions from the inside of the procedure of assembling the collective. But it also allows us to see how the concept of the collective *itself*, particularly in the way it demands ontological recognition of *all* existing associations, is grounded in political instead of modern/universalist/colonial ontology (i.e. the old Constitution).

However, Wynter herself aims to privilege a particular component of proposition 2 'sea abyss descendants': the Human. Drawing on Frantz Fanon, Wynter calls for a redescription of the Human by those whose autonomous existence has so far been denied. In the same way that 1960s feminists have extricated 'the Woman Question' from the Marxist frame, so should 'the Negro Question' become 'a question that is specific to [the] concerns' of those who own it (ibid., 312–3). 'Man' needs to be abolished and the 'Human' needs to rise (ibid., 261). But this privileging of the Human is in danger of establishing new hierarchies, which (re-)describe the Human once again in relation to what it is *not*. It leaves open the possibility of yet another ontological eradication of propositions that are defined as having no right to belong – an eradication that goes beyond the mere (temporary) exclusion from a particular collective for which Latour allows. In Wynter's account, there is little space for an autonomous external world; indeed, drawing on Aimé Césaire, she (ibid., 328) advocates the establishing of 'a new science' that studies nature 'condition[ed]' by our knowledge about our 'sociogenic principles'. In other words, the 'new science' needs to incorporate into its own self-understanding the limits of scientific theories in making decisions about the question of 'human natures' (Mignolo, 2009: 17, drawing on Fanon's terminology). This is in danger of leaving the two houses (of nature and of social relations) intact, and of granting the nonhuman world mere secondary status. It is in danger of (at least implicitly) once again signing up to the old Constitution and its modern understanding of what is 'real'. Although Wynter's call is based on the argument that 'the large-scale dilemmas that we must now confront as a species' – such as global warming – are in dire need of that new science, and of a new understanding of the Human, the 'species' that matters most is once again us. Nonhuman entities that are arguably heavily involved and invested in these dilemmas get little say.

How can we make space for a more fundamental challenge to modern ontology, which would involve giving nonhumans a greater say, without at the same time understanding ontology as *completely* autonomous from the historical background of the modernity/coloniality congruency? One way is articulated in/through proposition 2 'sea abyss descendants': it is the way of poetry. Glissant (1997: 8–9; emphasis added) gives both perished human and nonhuman elements of proposition 2 'sea abyss descendants' a poetic speech prosthesis that allows for their connection:

> For us, and without exception, and no matter how much distance we may keep, the abyss is also a projection of and a perspective into the unknown. Beyond its chasm we gamble on the unknown. We take sides in this game of the world. We hail a renewed Indies; we are for it. And for this Relation made of storms and profound moments of peace in which we may honor our boats. *This is why we stay with poetry.* And despite our consenting to all the indisputable technologies; despite seeing the political leap that must be managed, the horror of hunger and Ignorance, torture and massacre to be conquered, the full load of knowledge to be tamed, the weight of every piece of machinery that we shall finally control, and the exhausting flashes as we pass from one era to another – from forest to city, from story to computer – *at the bow there is still something we now share: [the] murmur, cloud or rain or peaceful smoke [to be found in and perceived from the abyss of the slave ship].* We know ourselves as part and as crowd, in an unknown that does not terrify. We cry our cry of poetry. Our boats are open, and we sail them for everyone.

Poetry is able to include both a historical trajectory and nonhumans (such as a cloud, rain, a smoke), without submitting to a common sense modern temporality or to fixed (scientific) laws of rationality. As Rolando Vásquez (in Ansems de Vries et al., 2017: 105) points out (and as the Glissand quote shows very well), poetics, in contrast to conventional understandings of science, is able to 'enter the thought of the unexpected, of what is in excess of the given'. Even Latour points out in one of his latest publications on understanding the connectivity of 'Gaia' how important it is to use language (and in particular poetry) in order to overcome the limits of modern (binary) thought (Latour, 2017: 71). But does that mean that science, due to the way it has been historically interwoven with the overrepresentation of Man, has no longer *any* role to play, and that poetry might be better able to tackle the great global challenges of the day?

Perplexity

According to Latour (2004: 104), one requirement for the assembling of the collective is the need to always remain open to the emergence of new associations that put propositions to the collective, leading to a non-stop asking of the question: 'How many are we?' (ibid., 108). For Latour, no one is better equipped to constantly challenge the existing collective with that question than scientists. Through their constant interaction with 'external reality', scientists bring into existence ever-new human-nonhuman-associations (though that does not mean that all of these associations need to be included) (ibid., 111). Through their experiments in their laboratories, their instruments, their models and their theories, the sciences cause *perplexity*: 'they make it possible to shift viewpoints constantly' (ibid., 137–8); they have great potential to 'stir up the collective' (ibid., 112).[3]

This was/is also the case in genetics. A decisive challenge to proposition 1 'survival machine' did not only come from the competing claim of proposition 2 'sea abyss descendants', but also from important scientific developments. Between 2003 and 2006, the Human Genome Project (HGP) set out to decipher and map the total number of genes in the human genome, and its results showed that there are far too few genes in the human genome to explain the vast amount of human organic traits, which hints at the more complex processes that are apparently going on in the interaction of genes, proteins and organic development. One discovery of the HGP was the existence of 'alternative splicing', which means that different genes can be and in fact are spliced together to generate the huge variety of proteins in the human body (Ho, 2010: 85). This confirmed a discovery in the late 1970s, when biologists found out that genetic coding sequences are not always continuous, but are often interrupted by non-coding 'introns', and spliced together after having been transcribed (ibid.). The ENCODE project that analysed 1 percent of the human genome in detail further subverted the concept of the gene as such, by showing that the great majority of genes are 'fragmented, intertwined with other genes, and scattered across the whole genome' (quoted in ibid, 2010: 85).

These experiments and discoveries potentially challenge the centrality of the gene in proposition 1 'survival machine'. How the gene precisely relates to the organism is no longer clear, let alone whether the result is indeed a 'survival machine'. The speech prosthesis that Dawkins has given to the gene breaks down, and genes speak up anew: 'We work differently than you, Dawkins, think we do'. There is no straightaway answer to the question of how they *do* work – we just don't really know. While the social sciences and humanities find it hard to leave the question 'How do you work?' hanging in the air (indeed, the question rarely comes up), it is their interaction with external nonhumans that leave scientists often no other choice than enduring that question, and enduring the silence that they might get as a response. They cannot define away the resistance of the material they deal with as easily as the humanities and social sciences can (cf. Disch, 2008: 92). In contrast to humans, nonhumans can consequently cause a perplexity that does not allow for an easy closing down of the question who and how many the collective is.

Proposition 3: 'emergent relationality'

> [I am] the relationship of the components . . . [I am] emergent and [my] causality [is] rarely linear or straightforward.
>
> Gilbert Gottlieb, quoted in Kathryn E. Hood et al.,
> *Developmental Science, Behavior, and Genetics* (2010): 4

Developmentalist biologists, who have opposed and been marginalised by the dominant gene-centric school of thought for decades, have taken the results of the HGP as confirmation of their competing theory of organic development. As Ho (2010: 86) argues, the HGP has proven that an organic system 'works by perfect intercommunication', diffusing 'the distinction between genetic and epigenetic, organism and environment.' The perplexity caused by the results of the HGP therefore results in a new proposition 3 'emergent relationality' that demands for proposition 1 'survival machine' to be excluded: instead of being a 'survival machine' fighting for resources in a hostile environment, it defines an organism as emergent relations in itself and with its environment. By grounding itself in a relational ontology and going against the 'survival of the fittest' argument, which has, as the previous witness has shown, resulted in the creation and justification of Man2, proposition 3 'emergent relationality' shows some affinity with proposition 2 'sea abyss descendants'. But proposition 1 'survival machine' would not go down without a fight.

Debate/struggle

Although the HGP has shattered the prospect of finding the grail to our existence in our DNA, the ideal of possessing the information to predict and control the development of life has not been given up. Despite the acknowledgement that 'predictability . . . as understood by a physicist or control engineer, remains an elusive goal', it 'continues to exert a powerful intellectual attraction' in the life sciences (Williams and Luo, 2010: 321; cf. Ansems de Vries and Rosenow, 2015: 1122). In this context it makes sense that despite the results of the HGP, many gene-centric concepts have not been changed, let alone abandoned (Neumann-Held and Rehmann-Sutter, 2006: 4).

One way of not giving up the ideal of control and clear-cut agency lies in the concept of 'information' itself. Although information is conceptualised as being ontologically separate from the process of its realisation – so that it can be thought about as generating and directing this process – it is perfectly able to disguise this role. 'Information' needs to be read, but at the same time animates this reading (or its translation); it literally 'in-forms' and is therefore a kind of subject without mind (Oyama, 2000: 14). Molecular biology grants information 'atomistic autonomy' that includes the capacity of both self-initiated movement and of being 'gathered, stored, imprinted, and translated' (ibid., 10). After the theory of 'gene action' had become out-dated, the metaphor of 'programme' became central in its reconceptualisation. The theory of the 'gene programme', which proposition 1 'survival machine' uses as well, continues to sustain the notion of genes as autonomous 'causal agents' while acknowledging at the same time the need for mechanisms of activation, without explicitly invoking the idea of purposeful creation (Fox Keller, 2000: 80). Such understanding opens the door for the possibility of 'reprogramming' by external intervention (ibid., 90). If the gene is a 'subject without mind' that itself needs activation, the 'mind/matter' dualism that distinguishes between (active) human subjects and (passive) matter remains intact. As subject, the gene continues to direct, while as object, it can be manipulated and modified, allowing for a substitution of 'Nature-as-designer' with 'Man-as-designer'. Man2 is not simply a purely biological being adapting more or less successfully to an environment characterised by scarce resources. Man, (still) endowed by God/Nature with the capacity for reason, is at the same time the designer who can *shape* His environment to His advantage. In Deleuze and Claire Parnet's words, the idea of design is not dispensed with 'even when it is accorded a maximum of immanence by plunging it into the depths of Nature' (Deleuze and Parnet, 2006: 68).

Another way of preserving gene-centric theory in molecular biology lies in the confining of 'complexity' to the level of how exactly genetic information is realised – in other words, to the detailed technicalities of the process of 'translation'. At the same time, a 'tacit reductionist commitment' to the Central Dogma persists at the level of general theory, even if it is not (always) explicitly acknowledged (Carolan, 2008: 757). A good example is this excerpt from a biology lecture about post-genomics and gene expression programmes, which maintains that

> [m]odern biology, being largely dominated by molecular and cell level approaches, has been criticised from several perspectives as being too gene-centric. Fundamentally though, it is difficult to argue that genes, and their protein products, do not play a central role in development.
>
> (Williams and Luo, 2010: 322)

Here, the centrality of the gene is taken for granted and the status of the protein is reduced to that of a gene 'product'. Moreover, it is argued that the Central Dogma 'is ultimately true' if it is correctly grasped as defining 'which information transfers are possible' (ibid., 324–5). It is suggested that Francis Crick himself formulated it correctly: 'once [genetic] information has got into a protein, it can't get back out again' (quoted in ibid., 324). With such a definition, it is easy for the authors to take current research results as a confirmation of the Dogma by pointing out that in 'its most recent form', the Dogma speaks of 'three possible types of information transfer': *general* (DNA to DNA, DNA to RNA, RNA to DNA), *special* (e.g. reverse transcription from RNA back to DNA), and *unknown* ('involving transfer of information in protein form back to nucleic acid form', which has 'never been observed in any living cell') (ibid.). The third type is particularly interesting, as it seems to indeed contradict the 'correct' definition of the Dogma by raising the idea that information *could* get out of the protein again; however, this option is more or less excluded by emphasising that it is only an unproven theoretical possibility.

Complexity is found at the level of the 'mechanics' of information transfer in a kind of 'black box' that does not touch the status and significance of 'the gene' as central carrier of information:

> The Central Dogma refers to the transfer of genetic information itself, and does . . . not mean that *on a mechanistic level*, the . . . function and activity of genes cannot be influenced by extra-genomic factors. . . . The *transfer of genetic information* should not be confused with the *mechanisms* by which these transfers are undertaken . . . the elegance of both the genetic code, and of the Central Dogma, belie the *immense complexity of the actual mechanisms* by which these information transfers are carried out in living cells . . . while the execution of the entire process itself is *stunningly deterministic*, there is no simple relationship between genes and outcome of their activity.
>
> (ibid., 324–6; emphasis added)

The identity of the gene is clear and comprehensible before we enter the 'black box' of its material realisation. Inside the box, something complex and potentially incomprehensible happens, before a 'product' pops out the identity of which we can once again clearly circumscribe. Complexity is of secondary relevance, insignificant for the generation of genetic theory. There might be some complexity going on between the gene, the protein and the organism, but it does not touch the fundamentals of proposition 1 'survival machine'.

Finally, capturing complexity and diffusion while sticking to traditional concepts of the gene is also enabled via pragmatically determining the identity of the latter in line with specific research needs. Indeed, as Paul Atkinson, Peter Glasner and Margaret Lock (2009: 4) point out, several studies have shown that the extent to which scientists among each other agree on the definition of the gene has been overestimated. For some, such as population geneticists, the precise definition is irrelevant, while others, for example genetic disease researchers, still operate with a narrow definition, despite being aware of 'real' genetic complexities (ibid.). Karola Stotz's study about the different concepts of genes employed by different biologists provides the insight that 'working biologists' mainly employ a 'consensus gene concept'. This is 'based on a collection of flexibly applied features of well established genes', which results in defining a particular stretch of DNA as 'a gene' if it has enough of these features. Stotz (2009: 235) concludes that this method 'inherently distracts from conceptually problematic cases', because most biologists divide a sequence in such a way that the resulting 'genes' match most closely what she calls the 'stereotype'. For example, although processes such as alternative splicing 'seriously undermine' the theory of a stringent relationship between genes and proteins, alternative splicing is commonly defined as *one* gene producing *multiple* proteins, instead of challenging the notion of the gene as such (ibid., 236). In this sense, the gene is indeed more 'fluid', as Ho (2003) and other developmentalist biologists maintain. But in mainstream genetics, rendering the notion of the gene fluid has become a pragmatic means of reinforcing its centrality – if the gene can be anywhere its identity and function are unchallengeable. Instead of acknowledging that the notion of the gene as an agent with a fixed identity might be an illusion, identity-boundaries are reinforced in a contingent, flexible way. While in Latour's collective, different scientists – population geneticists, genetic disease researchers, etc. – would give multiple and not necessarily commensurable speech prostheses to nonhumans, resulting in different propositions to the collective, the demand of the old Constitution to define external reality as *one*, ruled by *one* coherent set of laws, leads to the demand for *one* coherent theory of organic development. Man/gene remains unchallenged.

Man/gene's governance of the world

Proposition 4: 'controllable possession'

[I am the] industrial gene . . . that can be defined, owned, tracked, proven acceptably safe, proven to have uniform effect, sold and recalled.

Jack Heinemann, quoted in Denise Caruso, *The New York Times* (2007)

The biotech industry[4] gives the gene a speech prosthesis that is similar to the one that Dawkins uses – which is not surprising given that Man was reinscribed by Darwin 'as a purely biological being' at the same time as Man 'was redefined as optimally economic Man' (Wynter, 2003: 314). As Latour (2017: 77) points out as well, the idea of the 'selfish gene' is connected to 'a long line' of economic thinking that includes, among others, the theories of Locke and Smith. The concepts of scarcity and competition lie at the heart of both a modern capitalist market economy and mainstream genetics, and both go/went hand in hand with colonial and economic exploitation. However, while proposition 1 'survival machine' defines the gene mainly as an agent and driver, proposition 4 'controllable possession' defines it both as active agent and (more importantly) as passive product to be owned and sold. For the latter purpose, the gene claims to have discrete recognisable properties and clear boundaries; it is identifiable, differentiable and representable. It insists on stringent cause-and-effect relations between gene and organisms in order preserve its 'industrial' nature (Caruso, 2007).

Like proposition 1 'survival machine', proposition 4 'controllable possession' relies on the concept of a mechanistic world in which cause and effect can be clearly attributed to various organic or inorganic assembly parts. What is 'natural' needs to be clearly differentiable from what is manufactured, and complex organic developments need to be reducible to particular informational codes (Pellizzoni, 2011: 797–8). This condition seems to be satisfied with the concept of the distinct gene as material (and sole) carrier of information; the latter being translatable into a clearly recognisable organic trait (ibid., 798). But in the economic realm, the relationship between the material and informational identity of the gene is not straightforward. As material 'carrier', the gene is 'fixed' enough for it to be patentable. Being fixed in this particular way, it can also be labelled 'manufactured' as opposed to 'natural', which is important for patenting, as 'nature' cannot be owned. The informational identity of the gene, by contrast, needs to be diffuse enough to be traceable all over the final 'product' of the organism which, paradoxically, is then declared to be 'natural' to allow for easy regulation (cf. ibid; see also Calvert, 2007). Similarly to what happens in mainstream biology, the identity of the gene in proposition 4 'controllable possession' is fixed and fluid at the same time.

Debate/struggle

For Luigi Pellizzoni (ibid., 798–9), the ambiguity of the gene and its oscillation between different identity poles stands for how neoliberal governance works in general. Similarly to the biopolitics scholars that I have referred to in Chapter 1, he argues that governance is rendered flexible by grounding itself in an anti-essentialist, complexity-based ontology of biophysical matter (ibid.). But this argument overlooks how both the notion of governance as such and proposition 4 'controllable possession' confines complexity to an analysable field. Consequently complexity is always (necessarily) led back to representable (if shifting) identities.

In her study *Life as Surplus* (2008), Melinda Cooper analyses in detail the simultaneous complexity theory turn in both the life sciences and American neoliberal economic policy which, to her mind (ibid., 20), has led to the constitution of a 'new and mutable set of biopolitical relations' that has displaced the 'geopolitics of world imperialism' in the post-World War II era. Cooper (ibid., 44) briefly mentions that it was, however, only with the 'dissemination of mathematical models associated with complexity theory' – allowing the analysability (!) of complexity – that non-equilibrium ideas could 'attain a certain degree of credibility among economists'. But as Magda Fontana (2010: 591) points out, complexity theory-based modelling relies on nonlinear mathematics, which means that it offers no guarantee to be solvable. While linear mathematics, as exemplified in neoclassical economics, aims at and succeeds in finding general solutions to the problems it sets out, complexity theory can only provide 'particular explanations' which are less precise and rigorous (ibid., 592). This poses, as the following example will show, a drastic challenge to policy-oriented expertise.

For her argument about neoliberal theory Cooper mainly draws on the writings of Friedrich von Hayek, for whom governance in a neoliberal world needs to be in line with the workings of complex economic system that should not be interfered with. However, as scholars who are interested in the concrete governmental implementation of Hayekian neoliberal (or, as his supporters would call it, 'evolutionary') economics are well aware, this proves tricky on the ground of everyday policy-making. One (rare) compilation of studies that deals with questions of how to implement evolutionary economics is Kurt Dopfer's (2005) edited volume *Economics, Evolution and the State: the Governance of Complexity*. One of the contributors, Christian von Weizsäcker (2005: 43–4), argues that evolutionary economics can only seriously challenge the predominance of linear, function-based neoclassical economics if it manages to develop an 'evolutionary welfare economics'. This requires connecting 'positive statements' to 'normative statements', which means evaluating existing policies and drafting policy recommendations in relation to positive assumptions about economic processes. According to Wolfgang Kerber's analysis in the same volume, the difficulty of conceptualising positive policy from an evolutionary economics perspective has to do with the 'openness of economic processes and the resulting Hayekian knowledge problem' on which evolutionary reasoning is based (Kerber, 2005: 296). As von Weizsäcker (2005: 44) shows, positive policy formulation is crucially based on the measurement of value, for which in neoclassical economics the standard method is monetarisation.

As Kerber (2005: 296) stresses, attempts to combine governance and evolutionary economics are in need of a 'pragmatic compromise', and he consequently formulates three principles for evolutionary economic research (ibid; emphasis added):

1   Evolutionary economics should be pragmatic in a methodological sense, i.e. that in an ever-changing world *economic policy can and has to be made* despite the impossibility of eliminating all uncertainties in regard to its effects.
2   Evolutionary economics should not restrict itself to theoretical and basic research but should also do research about the application of evolutionary reasonings to the *solving of real-life problems* including participation in policy discussions.
3   For applying evolutionary arguments to policy questions we have to find a *pragmatic way* to *combine evolutionary with neoclassical arguments*, which to a certain degree will remain indispensable for many real-world problems.

As this makes clear, one strategy to uphold proposition 4 'controllable possession' against the challenge posed to it by the relational ontology in which proposition 3 'emergent relationality' is grounded, is to subordinate the notion of complexity to the pragmatic and common sense, even if this means leaving logic behind. 'Policy can and has to be made' in order to solve our 'real-world problems', but in order for that to work those 'real-world problems' have to be decoupled from the 'reality' of the natural systems that provided the basis for evolutionary/neoliberal economics in the first place. What about the power of these systems and their nonhuman participants to potentially undo and change our 'real-world problems'? This shows that an intrusion of nonhumans into the social house is only permitted as long as it does not threaten or make unmanageable the way the social house is generally ordered and governed. As Latour (2017: 78) argues, the fundamental element of understanding the earth as complex and all of its aspects (organic and non-organic) as inter-dependent – in other words, as 'Gaia' – is that 'externalities and internalities can no longer be easily distinguished'. This poses a fundamental problem for 'the neoliberal theory of action', because at its heart it continues to make this distinction, arguing that 'externalities . . . cannot be internalized by selfish individual agents' (ibid., 77–8). Neoliberals, Latour (ibid., 78) argues, 'are at war with the Whole – conceived as a State, or as Gaia' (ibid., 78). This goes along with a reinforcement of the division between the natural and the social house. Man/gene continues to govern the (social) world.

Dance of life

Proposition 5: 'dancing whole'

> [I am] a whole [performing] a highly coordinate molecular "dance of life" . . .
> [I am] layers upon layers of chaotic complexity [that] are coordinated . . . in mutual agreement in an incredibly elaborate, exquisite dance of life that dances itself freely and spontaneously into being.
>
> Institute for Science in Society, *Death of the Central Dogma* (2004)

Although the virtue of the sciences lies in the way they cause perplexity and stir up the collective, their problem is that they also confine reality to the actual, analysable field. For Deleuze and Guattari in *What Is Philosophy?* (1994), the objects of science are 'functions': independent variables characterised by 'distinct determinations that must be matched in a discursive formation with other determinations taken in extension' (ibid., 23). There are strong resemblances between what Deleuze and Guattari define as objects of philosophy – concepts that have no spatiotemporal, but only intensive coordinates, characterised by an inseparability of components (ibid., 20–1) – and how Deleuze describes the virtual field in *Difference and Repetition* (2004). The virtual field cannot be captured in or through scientific experiments and instruments. Instead it points us to that dimension of reality that can never be grasped and thereby never be controlled. The virtual field can only be *sensed* in singular experiences. It is precisely the attempt to communicate that sensation that we find in Mae-Wan Ho's scientific and political anti-GMO argument, which, as I maintain, is a good example for scientifically-grounded anti-GMO argument and activism more generally.

Proposition 5 'dancing whole' uses a description that is highly aesthetic and has the potential to reach its audience at a sensual level. Indeed, many critical and unconventional scientists, particularly those grounded in complexity science, go beyond mere scientific arguments and descriptions in order to make their case for what life *really* is about, using aesthetic and emotive language and images for that purpose: Coen, for example, confronts the common notion of the gene as alphabet or text with a visualisation according to which genes 'respond to "hidden colors"' (referred to and quoted in Fortun, 2009: 252), and James Lovelock (1990: 8) talks about the 'unceasing song of life' that is 'audible to anyone with a receiver, even from outside the Solar System'. Such strategies of dramatic 'aesthetisation' are strongly related to what Latour (2004: 63) calls the necessary transformation from 'matters of fact' to 'matters of concern': while a claim to 'facts' can only be made by a privileged few who have access to the acknowledged methods of establishing them, a 'concern' can be brought forward by all human-nonhuman-associations, and is to be debated by all associations. Determining what is 'real' is no longer solely up to scientists (narrowly conceived). The distinction between 'fact' and 'value' is dissolved, because a discussion of the latter presumes that a decision on the first is to be made elsewhere (cf. ibid., 95–102).

Ho would not go that far though. Indeed, identifying herself as a scientist, she also believes in factual truth and considers her own theory of how life develops and behaves as scientifically superior (see e.g. her book *The Rainbow and the Worm: The Physics of Organisms*, 1993). Nevertheless, she goes to great lengths to advocate a different understanding of science: one that overcomes the 'human chauvinism' that only allows for a concept of organic matter as passive object, and that aims to give what Latour would call 'nonhumans' a greater voice (Ho, 1993: 142, 100). According to Ho, there are no parts inside the organism – such as the gene – that can be singled out if we want to understand how life develops. How the organism works can only be understood in a holistic manner, which means that the traditional scientific approach of gaining knowledge via 'invasion' and breaking up a 'whole' should be replaced with a non-invasive method that lets the cell or even the organism as such 'tell its own story . . . to inform us of its internal processes' (ibid., 100; cf. Rosenow, 2012: 542). In that sense Ho might be quite happy to be part of what I have defined as proposition 5 'dancing whole'.

The witness testimony of the quantum physicist

For the quantum physicist, 'energy' and 'attraction' are the central principles of matter – which, as Ho argues, applies to organic as well as inorganic matter. In traditional physics, in line with the second law of thermodynamics, information is something gained by an active subject – a process that is energy-costly (Ho, 1993: 20). In the state of cost-neutral equilibrium, thermal energies can thereby not possess or convey information (ibid., 95). However, as the quantum physicist tells us, information 'is already supplied by the special structure or organization of the system' (ibid., 20). Due to electromagnetic attractive forces between molecules,

> an excitation of a specific frequency at a temperature where no other excitation of the same energy exists – i.e., in a system far from equilibrium – not only has just the requisite information to do the work, but the inherent power to do it as well.

(ibid., 20–1)

This understanding dispenses with the idea of separate entities and emphasises the significance of flows. The process of information-transmission is linked to electromagnetic forces (attraction) and excitation, and is no longer a static property to be found in an entity such as the gene. If a system is in a condition far from equilibrium, action is possible not because there is an agent directing and executing it, but because there is 'no other excitation of the same energy' (ibid.). In other words, material forces convey and 'decipher' information in a totally relational, non-causal manner.

Ho then takes this a step further by including the human observer into the immanent process of organic development. While, as previously pointed out, traditional genetics disguises the role of the gene as 'origin' and 'designer' by making it a kind of subject without mind, which preserves the mind/matter dualism between the human agent and the matter s/he controls, Ho (ibid., 142) maintains that the human observer can never stand outside of the system s/he investigates. Traditionally, the scientific observer is perceived as being 'strictly external to the system' and as not influencing the processes s/he observes, whereas in the quantum world, observer and observed 'seem somehow inextricably entangled' (ibid; cf. Rosenow, 2012: 542; Ansems de Vries and Rosenow, 2015: 1123). Observable particles can only become 'real' entities in the experiment: they are altered when scientifically manipulated and can only produce an effect, and thereby validate their existence, in the interaction with the one who manipulates them (Castelao-Lawless, 1995: 50). This process is captured in Ian Hacking's (1983: 23) famous phrase 'if you can spray them then they are real' (cf. Coleman/Rosenow in Ansems de Vries et al., 2017: 104). Consequently, the scientific 'observer' is immanent to the process of material reality that both human and nonhuman agents co-produce.

But in another passage of her book *The Rainbow and the Worm* Ho is then keen to emphasise that this state is only one of two, with the other one being utterly remote from direct human experience and influence, and therefore 'pure' (Ho, 1993: 144–5; emphasis added):

> [A]ll alternative possibilities open to the system co-exist in a *'pure state'* rather than a mixture of states until the instant we observe it. . . . A pure state is indivisible, it is a unity which we can represent as a 'superposition' of all the possible alternatives. The mixed state, however, is a mixture where the different states really exist in different proportions. The act of observation seems to put an end to this *almost dream-like pure state*, into one of the possibilities that previously existed only as a potential.

This description bears clear similarities to the way Deleuze distinguishes the virtual from the actual field (which also shows how much Deleuze himself was influenced by quantum physics). Although the 'pure' state is, according to Ho, no less real than the 'mixed' one, there is no way of directly accessing it without pulling it straightaway into the latter – indeed, the 'pure state' can only be 'dreamed' about (ibid.).

Later in her text Ho contemplates further ways of accessing this state of purity (ibid., 168; emphasis added):

Ideally, we ought to be one with the system so that the observer and observed become mutually transparent or coherent. For in such a pure, coherent state, the entropy is zero; and hence uncertainty and ignorance are both at a minimum. . . . It involves a consciousness that is delocalized and entangled with all of nature, when the awareness of the self is heightened precisely because self and other are simultaneously accessed. I believe this is the essence of *aesthetical* or *mystical experience*.

As previously mentioned, aesthetic metaphors and images are extensively used in quite dramatic ways by unconventional scientists to describe the function of the gene or the organism in the process of the development of life. In one of its reports, the Institute for Science in Society, which Ho has founded, uses the metaphor of 'dance' for this purpose – and this is where proposition 5 'dancing whole' emerges from:

> The best thing about the human genome project is to finally explode the myth of genetic determinism, revealing the layers of molecular complexity that transmit, interpret, and rewrite the genetic texts. These processes are precisely orchestrated and finely tuned by the organism as a whole in a highly coordinated molecular 'dance of life' that's necessary for survival. . . . All of this goes against the very grain of the Central Dogma that posits linear, mechanistic control. Instead, layers upon layers of chaotic complexity are coordinated, it seems, by mutual agreement in an incredibly elaborate, exquisite dance of life that dances itself freely and spontaneously into being.
>
> Institute for Science in Society (2004)

In a different text, Ho argues that the mechanic 'silent universe of lifeless, immobile objects' should be replaced with the concept of life as a 'vibrant world of colour and form, of light and music' (Ho, 1998: 76). The idea of life as being static, linear and mechanical is countered with a concept that stresses its vibrancy, art and non-reproducible originality. The idea of it being 'free' counters the idea of control, and the idea of it being 'spontaneous' the impossibility of planning due to a lack of choreography. Still, life is not 'chaotic', but immanently coordinated and regulated in a highly complex manner that cannot be (fully) understood. It is argued that the genome has 'a definitive "architecture" that holds up beneath the fluidity' (Institute for Science in Society, 2004; cf. Rosenow, 2012: 543; Ansems de Vries and Rosenow, 2015: 1123). Some of this testifies to the attempt to capture in a dramatised manner an understanding of (non-)human life that goes beyond what is straightforwardly explicable in scientific terms. It points at biophysical matter as that which has a force that can only be properly encountered at the level of 'sense'.

A different witness testimony

Arguments about chauvinism and the object needing to be able to tell its own story in a holistic manner are to some extent related to the thought of María Lugones

(2003) (cf. Chapter 4). For Lugones, oppression works through fragmentation: a fragmentation of the person as well as (related) wider societal fragmentation that sorts people into separate, homogenous, pure groups 'through unification and hierarchical ordering' (ibid., 141). Her call for a 'genuinely heterogeneous society' that has 'complex [!], nonfragmented persons as members' (ibid; cf. Vásquez in Ansems de Vries et al., 2017: 98) is strikingly similar to Ho's call for a recognition of the complexity of life – life that cannot be neatly separated into clearly-identifiable entities.

The difference between both strands of thought is the grounding of their claims. Lugones's reflections have emerged from an engagement with the experience of the oppressed, while Ho's are based in what she perceives as the (ahistorical) truth of nature, which leads to the latter at least implicitly upholding the old Constitution and its binary segmentation. One of the central tropes of anti-GMO activists has always been the description of GMOs as 'unnatural' or 'artificial', with one of the famous metaphors used for their rejection being the one of 'Frankenstein food' (see e.g. Jane Goodall, quoted in Poulter, 2015). By making a straight link between a manipulated, 'artificial' gene and the 'unnatural' or 'Frankenstein' organism, the centrality of the gene – as 'causing' the entire organism to be unnatural – is paradoxically reaffirmed, and the notion of the complexity or uncontrollability of life delimited. Arguing that it is possible to clearly define the identity of the GMO follows the same kind of black box logic that I have identified in relation to mainstream genetics: something that is 'known' (the modified gene) enters the black box of the organism, in which something complex, autonomous and partially incomprehensible happens, after which an end result pops out that can once again be clearly identified: 'unnatural' and 'monstrous' according to anti-GMO activists; 'natural' and 'safe' according to biotech proponents.

The professed desire to sustain or create definitive identities is generally neither surprising nor should it be outright rejected, given the political force that goes along with it. Indeed, it could be argued that it is the inability to differentiate and consider everything in flux that allows the kind of pragmatic interventions (e.g. the random shifting of the boundaries of the 'gene' that allow for a preservation of its identity as central developmental driver) that this chapter has so far identified as a crucial means to preserve the status quo in molecular biology and the governance of agricultural biotechnology. But it is not the determination per se that I find problematic: it is the *particular* identities that have emerged in much of anti-GMO activism. Our common sense, modern/colonial binary structures that do not allow for an understanding of complexity beyond its actual, analysable scientific dimension are not properly left behind – despite all attempts for transgression via aesthetic and other means.

Indeed, aesthetic, dramatic language is in danger of reinforcing binary thinking, particularly when it is used to describe the 'essence' of nature as one of cooperation, harmony and reconciliation. In a text that aims to celebrate the achievements of the Institute for Science in Society the late Member of Parliament Michael Meacher for example points out how new (complexity-based) scientific models have supposedly been able to reveal the 'beauty and harmony in nature' (Meacher, 2011: 8). Ho herself (2011: 9) defines 'the capacity to be inspired by beauty' (supposedly found in nature) as 'the fount, if not the *raison d'etre* of all creation.' And as Donna Haraway (1997: 60) points out, well-known environmentalist Jeremy Rifkin appeals in many of his publications and campaigns to the 'integrity of natural kinds' and the 'natural telos of the self-defining purpose of all life forms'. Instead of constituting identities that are more open towards non-binary difference, what is invoked is what Deleuze (2004: 64) calls, based on Hegel (though forwarding a somewhat different understanding), the 'beautiful soul', which maintains that that all differences are 'respectable, reconcilable or federative' (cf. Ansems de Vries and Rosenow, 2015: 1124; for a similar critique see Braun, 2002: 229). The 'beautiful soul' emerges from the wish for what Connolly (2017: 81–2) calls 'organic belonging', which expresses a 'sense of layered fit between self and world and between collectivity and world' that leaves no space for dissonance or the 'uncanny'.

Moreover, this sort of language and argument is also unaware how the creation of Nature as One historically went hand in hand with justifying the supremacy of White Man. And this is not just a thing of the past: Haraway (1997: 61) makes clear that the language of beautiful nature against monstrous artificiality is closely related to historical and contemporary racial and immigration discourses in Europe and the US that invoke a 'fear of the alien' and a 'suspicion of the mixed'; a fear of the 'impure' that, for Lugones (2003: 141), should be embraced by heterogeneous societies.

Her belief in the truth of nature makes Ho unable to overcome the ordering of reality along the axes of the house of nature and the house of social relations that the old Constitution prescribes – although Ho herself is convinced that this is precisely what she is able to achieve. Ho argues that if knowledge of how organic systems work were taken seriously, it would be used to guide us in the way we organise our own (human) systems. For her (1998: 273), the latter has to happen in a radically democratic way, because radically democratic systems are, similarly to complex organic systems, characterised by intercommunication, mutual responsiveness and the distribution of control. Ho (2010: 65–7) further maintains that such an understanding of how to organise society should replace our contemporary neo-Darwinistic 'survival of the fittest' model. Democracy, for Ho, is the most 'natural' form of both environmental and social organisation. Similarly, Stuart Kauffman (1995: 5, 28), who is one of the most distinguished and well-known complexity scientists in the popular realm, argues that 'the idea of a pluralistic democratic society' is not to be thought of as simply a human creation; it is 'part of the natural order of things'. In his account of life, Kauffman (ibid.) identifies 'hints of an apologia for a pluralistic society as the natural design for adaptive compromise' (cf. Rosenow, 2012: 539). The comparison between 'natural' and 'democratic' social systems reveals that both do not exist at equal footing, but that the 'truth' of the former is to be used as a levelling board for the latter. Ho's and Kauffman's attempts to draw parallels between the organisation of society and the organisation of nature is implicitly based on their very distinction – one has to 'learn' from the other; the 'natural' is primary, the producer of 'true' living, while society becomes nature's dualistic, inferior 'Other'. The (scientific) truth of nature as a harmonious democratic system is supposed to undermine less democratic forms of societal organisation. This can be more dangerous (in terms of a potential elimination of 'undemocratic' forms of life) than traditional political challenges to perceived democratic gaps, because 'truth' that is grounded in nature has few grounds upon which to be challenged (cf. Ansems de Vries and Rosenow, 2015: 1125). Moreover, because Ho's gaze is unable to travel beyond the supposed eternal truth of natural organisation, she is also unable to see how her notions of democracy and cooperation cannot be that easily de-linked from the 'imperial imaginary' (Mignolo, 2011: 48).

Having emerged from the perplexity caused by scientific experiments, instruments, and models, but not staying confined to that level, proposition 5 'dancing whole' overcomes the notion of 'fact' by attempting to reach out sensually in order to communicate a fuller experience of the reality of life. It thereby exposes the limits of rigid identification. By grounding itself, like proposition 3 'emergent relationality', in an emergent and relational ontology, it contests ideas of causality, agency, control, governance and ownership; and relatedly it also has the potential of shedding light on the oppressive and exploitative (colonial) dimensions of these ideas. However, due to its inability to de-unify and de-'beautify' Nature in a historical context, and to properly leave behind the idea of science being able to truthfully represent Nature as One, proposition 5 'dancing whole' in its current shape is still unable to listen to, consult and adequately acknowledge proposition 2 'sea abyss descendants'. While proposition 5 continues to appeal to an atemporal (natural) truth, proposition 2 grounds its claim for inclusion in historical trajectories of ontological and, more importantly, *literal* extermination.

In place of a conclusion: (un)making GMOs in the collective

As I have argued in the introduction of this chapter, a decolonised, politically ontologised collective provides a way to move beyond modern/colonial binaries that continue to shape discussions of science in relation to environmental activism. In this chapter, I have told a 'story' of the scientific GMO controversy in order to 'shrink' the universal, colonial claims of the old Constitution, and to also bring to the surface the history of modernity/coloniality in this specific context as an 'ontological conflict'. I have shown that coloniality does not only run through the Central Dogma of mainstream biology that continues, in parallel with the neoliberal governance of agricultural biotechnology, to shape the position that argues for the process of genetic manipulation being 'safe', but also grounds those anti-GMO positions that continue to locate Truth in a beautiful Nature about which we can find out via (reformed) scientific means.

But what does that mean concretely, for the way that environmental – and specifically anti-GMO activists – should politically move forward? To arrive at an answer to this question I now want to re-run the five propositions (though slightly modified/simplified) in the context of the procedural framework that

Politics of Nature sets out in Chapter 3, in which Latour (2004: 109) specifies four requirements:

> *First requirement*: **You shall not simplify the number of propositions to be taken into account in the discussion. (*perplexity*)**

'How many are we?' asks the collective.

Proposition 1: 'I am a machine, I exploit to survive. I am Man/gene.'

Proposition 2: 'I am the memory of murmur, rain, and smoke. I am abyss. I am original victim. I am poetry.'

Proposition 3: 'You don't know who I am. I don't exist but in relation to the whole.'

Proposition 4: 'I am always the same and my boundaries are set and inviolable. I can be owned. I can be controlled. I can be governed.'

Proposition 5: 'I am a highly-coordinated dancing whole.'

> *Second requirement*: **You shall make sure that the number of voices that participate in the articulation of propositions is not arbitrarily short-circuited. (*consultation*)**

One decisive prescription of the old Constitution is the need to short-circuit debates by referring to indisputable 'facts'. While the first requirement moves the collective away from this prescription by welcoming an ever-greater number of positively self-defined propositions into its midst, the second requirement already indicates that not every association can be included, but that every decision has to be preceded by a consultation that is lengthy and slow (ibid., 106). As I have already outlined in the introduction of this chapter, part of that consultation is the establishing of and listening to what Latour (ibid., 112) calls 'reliable witnesses' or 'spokespersons'.[5]

Throughout this chapter I have drawn upon a variety of 'witnesses' (or spokespersons) who spoke for or against the five propositions, and I have debated and judged their respective testimonies in relation to each other. At this point I want to analyse more systematically how a witness or spokesperson can be considered 'reliable' or (to use another term that Latour comes up with), 'faithful'. According to Latour, whether an act of spokespersonship is 'faithful' depends on whether those who/that are spoken for are 'allowed to make a difference in our thinking about them' (Latour, 1999: 117); in other words, whether they gain sufficient agency to modify other actors within the confines of the experimental protocol (Latour, 2004: 75). The idea of faithfulness as gaining agency to resist is related to Latour's general understanding of actors (human and nonhuman) being real if they have the power to have an effect, which Harman (2014: 45) defines as them being able to 'resist[ing] the trials of strength.' The decisive question is whether the way the experiment is 'staged' allows for that which is spoken for to redefine the problems it has been set; whether it can move beyond or outside of the space opened up by the questions it was asked, transforming the questions themselves. (Latour, 2004: 170–1; cf. Disch, 2008: 92)

Although I have brought Mae-Wan Ho as spokesperson and even as a contributor to particular propositions into the story of my collective, outside of this story it not quite that easy to define her as a 'faithful' witness for the human-nonhuman associations and propositions that come to the fore in the GMO controversy. As previously mentioned, the problem is that environmental activists often understand themselves far more as 'intermediaries' than as 'mediators': they attempt to speak for Nature as One and advocate an understanding of natural truth that is intrinsic and therefore atemporal (cf. Braun, 2002). Such understanding precisely leads to the sort of rigid binaries that go against the understanding of non-graspable and non-controllable complexity that many activists forward at the same time. Despite Ho's explicit attempts to overcome an understanding of science that divides the world into active subjects and passive objects, and despite her wish to 'upgrade' the object and to be willing to find out about it through mystery and aesthetics, the belief in the scientifically-discoverable truth of nature is not fully overcome. Complexity, as in more mainstream science, remains confined to the field of the analysable. One consequence is that anti-GMO activists are unable to productively dismantle status quo-preserving strategies that are used in pro-biotech conventional science, in the biotech industry and in governmental regulation. Moreover, the old Constitution and its decisive feature of dividing the world into two houses remain intact, regardless of all attempts to dismantle it.

If anti-GMO activists such as Ho were to turn from being 'intermediaries' to being 'spokespersons' or 'mediators' in the way Latour envisages, they would have to change their position to deliberately 'stage' an act of representation in what Latour (2004: 147) calls a particular political 'scenario' (which would be the equivalent to the scientific experiment). In this particular 'scenario' the gene, or the organism, is conferred agency in the sense that it should be given the chance to reveal itself and make a difference. Does it gain sufficient agency to *resist* or *redefine* the identity with which it is endowed? And what kind of impact would that have on the ones who 'staged' the political scenario in the first place?

I now want to give one example of what could be considered 'faithful' spokespersonship in the 'real world' of the politics of agricultural biotechnology. This example has emerged from my analysis of the Word Trade Organisation (WTO) trade dispute between the US/Canada/Argentina as complaining and the European Communities (EC) as defending party about the alleged EC ban of the import of GMOs between 2003 and 2006 (the case of *EC Biotech*) (cf. Rosenow, 2009). Much of the dispute centred on the question whether or not it can be scientifically proven that GMOs are safe to plant and consume, because the WTO Sanitary and Phytosanitary Agreement gives a WTO member state the right to ban imports if there are legitimate concerns about adverse health or environmental effects. It was for this reason that scientific experts were invited to give evidence before the WTO *EC Biotech* Dispute Panel.

The US, Canadian and Argentinian delegations were keen to interrogate these experts on questions of the risk/safety of the disputed GMOs by asking specifically about the latters' 'properties'. Any problems that the experts mentioned in relation to the wider environmental impact of GMOs got the response from the delegations that these concerns were not directly linkable to the GM *method* (as that which created the particular GM *product*), but merely to the environmental *effect* generated by the technology, and were thereby irrelevant for the *EC Biotech* dispute. For example, a genetic modification that makes a plant resistant to a certain herbicide (indeed, most genetic modifications are done with the aim of either reducing the impact of herbicides on the plant, or making the plant resistant to insect pests; Macnaghten, Carro-Ripalda and Burity, 2015: 14) naturally leads to an increase in the use of that herbicide – an effect that can be questioned, but, according to the complaining parties, not on the basis of how it was achieved. As the Canadian delegation emphasised (World Trade Organization, 2006: DS291(2,3)R, 4.1041; emphasis added):

> Drs Snow, Squire and Andow agreed that, in principle, the ecological effects of herbicide tolerant cropping are similar regardless of whether the HT crop was developed through transgenesis or mutagenesis. So apparently *apples are in fact apples after all and not pears.*

Although the experts that the Canadian delegation referred to indeed agreed on this 'in principle', they did not make the same clear distinction between cause and effect – for them, it was the overall impact of the new technology that mattered:

> it is also clear that the reason you use the GMHT plant is so that you can use the herbicide. It seems from a scientific perspective that the issues like resistance in weeds are scientific issues to be considered with respect to GMHT plants.
>
> (Dr David Andow in ibid, Annex H: 1196)

Now this is interesting when thinking about science and spokespersonship. In above quote, Andow feels compelled to answer from a 'scientific perspective', because this is the role with which he is endowed in the dispute case. From a more traditional scientific point of view, however, it could very well be argued that his assessment here is not properly scientific. After all, he was not asked to provide an analysis of the impact of modern agriculture on wider ecosystems. Instead, he was asked to determine, in line with conventional risk assessment procedures, whether the genetic modification of an organism *as such* directly poses a risk – which all of the scientific experts (including Andow) denied. But Andow insists on arguing *as a scientist* that the indirect impact on the environment matters as well.

In this particular example Andow (against his contention) did *not* answer the question he was asked. Instead he translated the question into a different one that he felt – as a scientist – he should, and could, answer instead. For him, what needs to be considered when talking about 'risk' or 'safety' necessitates looking at modern agriculture as a whole:

> With the food/feed side there is generally a presumption that the presently consumed food is safe. Whereas with the environment, the presumption

that the current production systems are safe to the environment is no longer granted. Indeed the current varieties produced by conventional methods are no longer all presumed to be safe. . . . There has been an evolution of concerns associated with agricultural production, that stems over a period of about fifty years. To some extent, GM crops are swept up in this broader movement of the evolution of concerns of new agricultural technologies that has been changing over a longer period of time.

(ibid, Annex H: 1194)

Following this trail of thought, risk assessment procedures need to be transformed for the whole of agriculture, and not just for GM-products. This questions not just the US/Canadian/Argentinian position, but also the EC approach to agricultural biotechnology, insofar as the latter based its import ban solely on the dangers supposedly posed by the end product of GE technology. In line with the arguments put forward by many anti-GMO activists (who indeed supported the EC in the dispute), it assumes that conventionally bred organisms are different from their genetically modified counterparts. Today, it therefore defines its task as creating transparency about whether a product is genetically modified or not, both via tracking-/labelling regulations and via guaranteeing the 'coexistence' of transgenic, conventional and organic agriculture (Lezaun, 2006: 1). Andow's position leads to questioning the need to neatly distinguish between GM and conventional agriculture, which, after all, might not lead to any of the more fundamental changes in agriculture that Andow deems necessary. It also leads to question the normatively loaded distinction between the 'natural' and the 'unnatural' that is so prevalent in anti-GMO activism – although it needs to be stressed that most anti-GMO motivated activists also challenge a wide range of conventional forms of agriculture, emphasising the need for sustainability, local production and decentralisation (see e.g. Open Letter from World Scientists to all Governments, 2000; Pretty, 1998; cf. Chapter 3). But still, the binary between 'GMO' and 'natural product' continues to feature highly among activists, and most campaigns advocate either the banning or the tight regulation of GMOs – also because this feeds into a straightforward, practical policy-demand and thereby constitutes a viable campaigning strategy.

It could be argued that in this particular 'political scenario' at the WTO, Andow acted as a spokesperson in the way Latour envisages. The object that Andow confers agency to has been able to shift the very question it was asked, from the risk associated with singular products to the risk that the impact of wider technologies poses to wider ecosystems. His translation was characterised by making non-equivalent situations equivalent – which, according to Latour (1983: 155), is typical for laboratory science (which 'translates' real-life natural problems into laboratory problems). With a twist on the WTO Canadian delegation's proverb, it can be argued that laboratory science thereby indeed regularly translates 'apples' into 'pears'. This is insofar interesting that it is neither recognised nor accepted by those scientists, environmental activists and policy-makers who demand that 'sound' science ('facts', in other words) be a central criterion for proper political decision-making.

Relating this to the 'story' of this chapter, Andow's spokespersonship could be taken as advocacy for proposition 3 'emergent relationality' and proposition 5 'dancing whole': 'You don't know who I am. I don't exist but in relation to the whole [. . .] I am a highly-coordinated dancing whole.' Or, to put in Latour's words: 'No one knows what an environment can do' (Latour, 2004: 80).

> ***Third requirement*: You shall discuss the compatibility of new propositions with those which are already instituted, in such a way as to maintain them all in the same common world that will give them their legitimate place. (*hierarchization*)**

It should be clear by now that propositions 3 'emergent relationality' and 5 'dancing whole' are incompatible with propositions 1 'survival machine' and 4 'controllable possession'. Indeed, Andow's lack of concern about apples being pears or not is challenging central ideas of trade and ownership. His analysis diffuses the idea of stringent cause-and-effect relations, and thereby implicitly also questions the way that trade can take place and be regulated. After all the very idea of trade and its regulation is based on the notion of discrete tradeable goods/products á la proposition 4 'controllable possession', with notions of 'risk' being attached to those products. But what Andow did was to translate concerns about GMOs into concerns about ecosystems, no matter from where the damage that is caused to these systems is coming. Relatedly, giving up the notion of stringent cause and effect also goes against the 'survival machine' speech prosthesis.

But what about proposition 2 'sea abyss descendants'? To what extent is that proposition compatible with the other propositions? Acknowledging this proposition is in need of reflecting, as explained throughout this chapter, on historical developments and historical relations of oppression; on how the right for ontological self-definition that Latour values so highly might give those whose existence has been historically written out and defined as merely one of 'lack' priority when it comes to competing claims for inclusion into the collective. There are strong resemblances between proposition 2 'sea abyss descendants' and propositions 3 and 5, particularly in relation to the need to forward a relational ontology and politics (cf. Vásquez in Ansems de Vries et al., 2017: 97), to challenge fragmentation, and to let what-/whoever has been excluded define him/her/itself anew

Although both sets of propositions continue to have problems with each other, there is no reason why coexistence in the collective should not be possible. What is crucial for Latour in his definition of the collective is the concept of the *experiment*. The collective does experiment with the 'attachments and detachments that are going to allow it, at a given moment, to identify the candidates for common existence, and to decide whether those candidates can be situated within the collective' (Latour, 2004: 196). Towards the end of *Politics of Nature*, Latour (ibid., 216) introduces the figure of the diplomat, who helps propositions in this process:

> "At the bottom," [the diplomat] says to them, "you don't know . . . what you were holding to before I got the negotiation going. You have just discovered

how much you care about this treasure; you would perhaps be prepared, then, to house it in a different metaphysics, if by doing so you could increase the size of the common house. Would you be ready to shelter those whom you took to be enemies but who have just taught you what you cherish more than anything in the world?"

Would the propositions that emerge from or are related to anti-GMO activism be prepared to let go of the metaphysics of the Truth of Nature (which continues to be linked to coloniality) in order to give better expression to the different ontologies that they are trying to communicate? Would anti-GMO activists be prepared to, in exchange for that, accept a proposition that potentially demands the giving up of the idea of beauty, telos and holism of Nature – and the idea that the GMO is 'unnatural'?

Running with Andow's spokesperson worries about risks to ecosystems, but relating them to wider societal and political concerns, one potential political consequence could be the call for activists to focus far stronger on 'the struggles of farmers, NGOs, and researchers committed to producing food in more sustainable and socially just ways' instead of buying into the easy natural/unnatural binary – even if that might cost them the 'popular support' that is easily gained 'by mobilizing fears' (Schurman and Munro, 2010: 153) As Schurman and Munro (ibid., 182) point out, focusing on these struggles enables us to understand that 'the challenges facing farmers and the poor have complex socioeconomic, political and institutional [and, as I want to add, *historical/colonial*] roots and therefore require a variety of strategies'. As decolonial scholars stress, particular attention needs to be paid to the struggles and claims of those who are located 'outside' of Europe or the US, in different geo-historical, colonised places (Mignolo, 2007: 458). This is a perspective to which the next two chapters will turn.

> ***Fourth requirement*: Once the propositions have been instituted, you shall no longer question their legitimate presence at the heart of collective life. (*institution*)**

The consultation of the collective is closed – for now.

Notes

1 According to Harman (2014: 12) it is only in the latest stage of Latour's work, particularly his recent book *An Inquiry into Modes of Existence: An Anthropology of the Moderns* (2013) that Latour finally moves away from ontologising politics.
2 The ontological conflict that my story enacts seems to forward the understanding there is *one* ontology that is in conflict with *another*. Such an approach can obviously be critiqued for precisely brushing over the multiplicity of ontologies, and for yet again reinforcing binary thinking. However, as Blaser (2013: 553) points out, it is 'one of political ontology's concerns . . . how to operate in a terrain dominated by conceptions of an all-encompassing modernity', which means that the principle objective becomes the highlighting of the existence of something other. In other words, the main aim for Blaser (and for my chapter, and even my book more generally) becomes the making

space for the 'other(s)' in general, rather than the fleshing out of what 'other' ontologies concretely entail.

3 Although Latour obviously means 'the sciences' here in a narrow sense, this does not need to be the case. What it means to engage in 'science' can be understood in different ways – it can include, for example, the way Indigenous people have been involved in 'experiments' with the nonhuman world for centuries (cf. Tilley, 2017). Due to the transitional nature of this book, a more in-depth engagement with these 'other' ways of doing 'science' has not been possible at this point.

4 I refer to the biotech industry and not Heinemann as giving the gene a speech prosthesis, because Heinemann describes rather than supports proposition 4.

5 Here I use both terms interchangeably, because *Politics of Nature* itself does not make a proper distinction – though the term 'spokesperson' is often used to talk about the way nonhumans are spoken for by scientists.

3 Voices and visibilities

The Indian Bt cotton controversy

> The Indian seed industry is rapidly moving into a phase of 'corporate control over seeds' with the introduction of transgenic crops . . . [W]hat begins is war with Indian agriculture. What begins is the destruction of Indian agricultural diversity. What begins is the dependence of Indian farmers on industrialised, unsustainable techniques of the developed nations. . . . What begins is the launch of neo-imperialism of seed and food.
>
> Vandana Shiva, Ashok Emani and Afsar H. Jafri (1999)

As I have indicated in Chapter 1, resistance to agricultural biotechnology in the so-called Global South is often related to a broader critique of the consequences of neoliberal globalisation and neo-colonialism. Many anti-GMO activists – such as prominent Indian writer and activist Vandana Shiva in above-cited quote (Shiva, Emani and Jafri, 1999: 601–2) – regard as interrelated the planting of GM crops and the power of global corporations to promote their technologies. Moreover, they see these developments as intrinsically related to past colonial exploitation. This perception relates to the decolonial argument that contemporary expressions of globalisation can only be fully understood by reference to the global character of the modernity/coloniality project that has always relied on ' "subalterniz[ing]" other local histories' (Escobar, 2007: 183). Shiva and other Global South environmental activists might agree with Mignolo (2007) when he argues that emancipation – or rather, considering the problems of the term from a decolonial perspective, *liberation* – can only be found by moving 'outside' of Europe and the US (ibid., 458). For Mignolo (ibid.) we need to prioritise those understandings that come 'from a country geo-historically located beyond both'. As Escobar (2007: 186) maintains, the term 'outside' often creates misunderstandings:

> In no way should this exteriority be thought about as a pure outside, untouched by the modern. The notion of exteriority does not entail an ontological outside; it refers to an outside that is precisely constituted as difference by a hegemonic discourse . . . By appealing from the exteriority in which s/he is located, the Other becomes the original source of an ethical discourse vis à vis a hegemonic totality.

As this chapter will show, a mere surface analysis of Shiva's arguments about agricultural biotechnology would lead to her being squarely placed in the thought from the 'outside' that decolonial scholars envisage. Indeed, Mignolo himself (2010) embraces Shiva's work, based on the argument that her 'analytic of globalization and the horizon for future non-imperial societies' is informed by 'de-colonial thinking' (ibid., 18–19). Indeed, Shiva (2009: 18) argues that agricultural biotechnology relies on a patriarchal-colonial logic that embraces mastery and exploitation, and that stands in opposition to 'the feminine principle', which has supposedly been practised by women in the Global South in pre-colonial times (ibid., 18, 31). However, Shiva also grounds her opposition to GMOs in scientific reasoning (Shiva, 2015), and she ignores the agency and voice of those Indian farmers who are in support of GE technology.

This chapter will argue that in order to come to terms with the ambiguity of Shiva's argument and standpoint, a *decolonial* perspective should be complemented with a *postcolonial* one. Indeed, as I will show, it is useful to return to Spivak's famous 1988 argument about the (lack of the) voice of the subaltern. To what extent is at least some of the decolonial critique and analysis based in the understanding that the subaltern *can* speak for her/himself? To what extent is this related to a lack of in-depth analysis of specific colonial trajectories in specific regions of the world (particularly outside of Latin America)? And to what extent is an emphasis on thought from the 'outside' in danger of brushing over the problem of representation that Spivak so aptly points out? Spivak's nudge to develop an attitude towards the subaltern that goes beyond both the (liberal) affirmation and the (critical) outright rejection of traditional notions of representation is in many ways related to my concern about the need to move beyond binaries in understanding and conceptualising radical environmental politics. Moreover, as I will show in this chapter, it can also be linked to Latour's call for a politics of representation in which humans start acting as spokespersons. Indeed, linking Latour's critique to Spivak's enables us, I maintain, to find a new approach to not only the voice of the nonhuman, but also the voice of the silent (subaltern) human. Such approach requires both a reflection on differentiated geopolitical positions (as Spivak argues) and a proper understanding of the need for a radically different ontology that is able to dismantle the modern (and, I argue, colonial) two-house logic that both Latour and, by implication, decolonial scholars are concerned about.

The chapter will start with a detailed investigation of the claims about farmers 'voices' and opinions that we can find in Shiva's argument and in the argument of her principal, pro-biotech opponent – Ronald J. Herring – about the introduction of the first and most significant GMO in India: Bt cotton. The first section will be particularly interested in the binary frameworks that underlie both Shiva's and Herring's analysis. I will then relate this discussion to the problem of the very concept of 'voice', through an analysis of Spivak's famous argument in conversation with Latour's emphasis on the need for spokespersonship. The chapter will point out the problems that occur when farmers' opinions are sampled through a focus on what they literally say, rather than an analysis of what structures and conditions their 'voice'. This argument will then be enriched through an elaboration on Deleuze's metaphysics of transformation, particularly by looking at the distinction that Deleuze makes between the 'sayable' and the 'statement', between 'perception' and 'visibility', and how he and Guattari conceptualise 'regimes of signs'. Using the different ontology that emerges out of Deleuze's metaphysics in dialogue with decolonial critique, the chapter will provide a fuller understanding of what is at stake in the Indian Bt cotton controversy, concluding with some concrete suggestions about how to move forward in anti-GMO activism. I will advocate an approach that calls for the need to endure the discomfort caused by the silence of subaltern-nonhuman voices in the Bt cotton controversy, and to embrace the disruption that this silence causes the 'feel good' understanding that many environmental activists have of nature and their own position in the world. At the same time we need to embrace the 'noise' that accompanies what I will call, based on Deleuze, the creation of new 'statements' in human-nonhuman-(anti-)GMO assemblages. These assemblages and statements have the power to break up existing striations of power and domination, but often turn too quickly to making new, rigid, colonial striations in relation to questions of identity, nature and the 'right' activist way.

Who's speaking? Indian smallholders and Bt cotton

Assessing the success or failure of Bt cotton in India has become crucial for debates around the need for agricultural biotechnology in the Global South. As Glenn Stone and Andreas Flachs (2014) maintain, when it comes to questions of hunger and poverty Indian Bt cotton is and was one of the 'most closely watched and hotly debated cases of GM crops' (ibid., 649), because it 'is the one GM crop that is widely planted by smallholders' (ibid., 652). Bt (Bacillus thuringiensis) cotton, a genetically modified cotton that contains a toxin-producing bacterium gene coding that repels pests such as cotton bollworms, was officially introduced into India by a joint venture of multinational agricultural corporation Monsanto and Indian company Mahyco in 2002. Its legalisation was preceded by a few years of field trials as well as the illegal hybridisation and dissemination of seeds on the black market (cf. Shah, 2008). In 2013 an indefinite moratorium was placed on all Indian GM crops and field trials with the exception of Bt cotton, which at that point had almost universally been adopted. The moratorium resulted from the discovery that a positive expert report that had made possible the commercial release of Bt brinjal (a GM aubergine variety) turned out to having been plagiarised from a pro-biotech lobbying document. However, in 2014 approval was again granted for conducting open field trials of 14 food crops (Egorova, Raina and Mantuong, 2015: 108).

The battle against the introduction of Bt cotton in India started in 1998 with the arrival of the first field trials. The Karnataka Rajya Raitha Sangha (Karnataka State Farmers Organisation – KRRS), led by prominent activist M. D. Nandujaswami, was at the forefront of campaigning against GMO crops and related issues. It became the centre of the NGO-led campaign 'Monsanto leaves India' under the umbrella of which a range of sub-campaigns were launched, such as the campaign

'Operation Cremate Monsanto' that focused on the burning down of Bt cotton field trials in Karnataka (Nanjundaswamy, 2003; cf. Egorova, Raina and Mantuong, 2015: 107). 'Operation Cremate Monsanto' was backed by international NGOs such as Greenpeace and global grassroot movements such as Vía Campesina, and it was very much in tune with other alter-globalisation movements of that time (Herring, 2006: 468–9). Shiva, who also supported the campaign, has been one of the most prominent figures among those who have been arguing and campaigning against Bt cotton ever since. Shiva has coined the famous term 'seeds of suicide' (2007: 145, 2000), which alludes to an increase in the number of suicides of Indian farmers in the early 2000s. While government officials attributed this increase to natural failures (e.g. failures of the Monsoon), poor individual management and alcoholism (Madsen, 2001: 3738), Shiva and other anti-GMO activists blamed corporations such as Monsanto as well as the 'unholy trinity' of the WTO, the World Bank, and the International Monetary Fund (IMF) (Shiva et al., eds, 2000)

One of the most prominent critics of the campaign (and of anti-GMO activism in India more generally) is Ronald J. Herring, who alleges that GMO critics deliberately turn a blind eye to GMO-supporting facts such as the speed with which Bt cotton has been picked up by farmers due to the higher yields it brings. Moreover, anti-GMO activists also continue, Herring (2006) maintains, to spread myths such as the ones about GMOs being responsible for farmer suicides as well as the 'terminator hoax' (which is the claim that Bt cotton seeds are sterile and have to be re-purchased every year). For Herring and N. Chandrasekhara Rao (2012), there is sufficient evidence for the argument that biotechnology provides a way out of the 'classic poverty trap in agriculture: low yields, low income, low investment, resulting in low yields and on to another cycle' (ibid., 46). Those campaigning against GMOs are mostly, Herring (2006: 484, 489) argues, 'metropolitan elites' who fail to adequately understand the reality of the rural poor, or, as he puts it, 'the brute compulsion of economic facts'.

Both the pro- and the anti-Bt cotton narratives feature a strong appeal to the 'facts' of the situation that is debated, and both claim to have 'truth' on their side. For Shiva, support of GMOs is identical to support of neoliberal globalisation and the accompanying privileging of corporate profit over public welfare: she and her co-authors maintain that the benefits of biotechnology go exclusively 'to the seed and chemical corporations', while 'the costs and risks are exclusively born by the small farmers and landless peasants' (Shiva, Emani and Jafri, 1999: 601). This argument is closely tied to the one about neo-colonialism: for Shiva, Emani and Jafri (ibid.), the agents of globalisation and agricultural biotechnology are firmly located in 'the first world', providing 'technologies and investment', while the recipients in the third world (whether farmers or national industry) obey the commands of the former 'without understanding and assimilating the inherent adverse impacts of genetically modified crops.' As already mentioned in the introduction of this chapter, Shiva also brings an eco-feminist perspective into the debate: for her agricultural biotechnology is part and parcel of a patriarchal-colonial take on nature that relies on mastery and exploitation. This is diametrically opposed to what she calls 'the feminine principle', which, having been traditionally practiced by women in the Global South, is supposedly oriented towards the 'small'/local, based on respect for integrity and diversity, and on avoidance of competition and aggression (Shiva, 2009: 18, 31). Shiva sees the solution to the problem of poverty in the related vision of the 'Village Republic' that originates in Ghandian philosophy. Going against both centrist state and global governance regulation, it advocates the establishment of a governance structure that is based on decentralisation, participation, autonomy and direct democracy, with an economic system that is characterised by sustainable small-scale agriculture, resulting in a life in harmony with nature (cf. Nanjundaswamy, 2003). Local, apparently self-determined ways of life that are supposedly embedded in Indigenous culture and tradition are celebrated and contrasted to the hierarchical and hegemonic features of a world that is dominated by global corporations.

For Herring (2006), this account is epistemically violent because it is based on a fantasy of farming that is not only distant from the daily hardship of farmers in the Global South, but also fails to take seriously the agency that farmers show when they adopt GM technology. Indeed, Shiva, Emani and Jafri (1999: 603) suggest that all farmers who have come to the conclusion that Bt cotton has been successful have been manipulated by the 'propaganda' of the 'biotechnology lobby', large, well-off farmers or seed distributors who work for the seed companies (cf. Shiva, 2015). Herring (2006: 484) counters that it is easy to make such claims in a context where the ones spoken *for* (particularly the rural poor) rarely get the chance of having their *own* voice heard by a wider audience. He defines the position of anti-GMO well-known activists and writers such as Shiva as one of epistemic brokerage that is ignorant of what is really happening on the ground (Herring, 2009). Drawing on his and Rao's interviews with farmers in Warangal district of Andhra Pradesh (Herring and Rao, 2012), interviews with farmers in Gujarat (Roy, Herring and Geisler, 2007), statistics about the increase of yield when Bt cotton has been planted, and the simple fact of now almost universal adoption of Bt cotton in India, Herring aims to prove these brokers wrong: farmers, he maintains, are rational agents whose 'attitude is essentially empirical', being determined by 'what works' (ibid., 171).

What Herring's critique of Shiva and other anti-GMO 'epistemic brokers' neglects is that the framework from within which he operates – one that affirms Enlightenment notions of (technological) progress and rationality – deprives farmers of actual agency and voice as much as Shiva's account (cf. Rosenow, 2013). While Shiva is only able to see agency if it confirms her narrative of nature, Indigeneity and the workings of global power (in other words, if farmers *resist* the adoption of biotechnology), agency only falls into Herring's view if farmers adopt and support GMOs. For both Shiva and Herring, the practices of those who do not cohere with their respective accounts are only explicable by pointing at manipulation. While Shiva is indeed in danger of over-emphasising and over-simplifying the power of multinational corporations and other dominating 'outside forces', Herring's analysis, by contrast, lacks any systematic analysis of power. One particularly striking example for the consequences of this analytical blind spot is when he seems to take at face value the (liberal) claim that '[i]n a market economy, it is difficult to conjure the mechanisms through which dictatorial or coercive powers could be imposed on a million of individual choices about cotton-seeds' (Herring, 2006: 472). Herring makes ample use of seemingly 'radical' terminology, by talking about class differences and material interests. But this terminology is only invoked in order to contest the assumed power of the 'urban', 'post-modern', 'cultural studies'-influenced left-wing 'elites' that fail to understand what really matters to the peasant class, and not in order to contest the forces of capital (see e.g. Herring, 2009). All of this results in a failure to take into account structures of domination that have the potential to *shape* the voices and behaviour of farmers that he bases his analysis on:

> [W]hy would farmers assume the risk of illegal activities [such as circulating illegally hybridised Bt cotton seeds] to put themselves and their families on a suicidal path? Does global diffusion of agricultural biotechnology indicate false consciousness on the part of farmers? Are they duped? Innumerate? Incapable of learning?
>
> (Herring, 2009)

Despite his class rhetoric, it seems to be inconceivable for Herring that there might be forces at play – economic, cultural, historical – that are outside of the grasp of the individuals who are subjected to them. By contrast, Shiva has a very clear (if over-simplified) theory of these forces: 'Within the European concept of property', she argues, 'capital is the only kind of investment with value' (Shiva, 2006: 24).

For Glenn Stone (2012), Herring's approach is blind to the politics of science. While Stone (ibid., 63) agrees with Herring's critique of the anti-GMO 'reciprocal NGO authentication system' in which truth is primarily determined by what fits the already-existing narrative, he alleges Herring of ignoring the 'industry-journal authentication system'. The 'truth' of science is determined by its verification in academic journals, which often publish findings (particularly in relation to biotechnology) that constitute straightforward, easy-to-understand results, providing the kind of 'empirical impact narratives' desired by an industry that is keen to provide positive assessments of its field performance (ibid., 67). The industry provides scholars with otherwise relatively inaccessible data and funding for their research, and it provides journals with publicity. A more careful analysis of a potentially complex situation 'would both slow down the pace of publication and make for much more qualified conclusions, resulting in fewer publications, in less selective journals, with lower impact' and would thereby go against the interests of all involved stakeholders (ibid.).

Stone (ibid., 65–67) provides a convincing account of various selection and cultivation biases that have influenced the results of the researchers (mostly economists) that Herring and Rao cite in their 2012 article about the benefits of the introduction and adoption of Bt cotton in India (cf. Stone and Flachs, 2014). In a different article, Stone (2015) also draws on his own research in Warangal district of Andhra Pradesh to show the extent to which Herring's field research, with its narrow focus on individual 'voice' and 'behaviour', fails to adequately grasp the historical, cultural and social complexity of the situation of Indian smallholder agriculture. In particular, Herring, according to Stone, is unable to take into account the systematic 'deskilling' that has been taking place among India's smallholder farmers for several decades. Deskilling goes back to social complexities 'unlikely to be included in voice studies', and it is indeed even 'actively discounted by soliciting individual farmer opinions' (Stone and Flachs, 2014: 651). This argument resonates with Esha Shah's (2008) anthropological and historical study of resource-rich Bt cotton farmers in Gujarat, which comes to the conclusion that the embrace of Bt cotton has mainly to do with a specific 'value framework' that has become deeply entrenched in Indian agricultural practices since the beginning of the (state-driven) Green Revolution in the mid-twentieth century. This value framework is characterised by a strong belief in never-ending scientific/technological progress and is resistant to 'paradigmatic changes' (ibid., 433). Contextualising his analysis with reference to existing cultural and social systems in Gujarat that determine access to land, water and (cheap) labour, Shah's conclusion directly contradicts Herring's argument (ibid., 444; emphasis added):

> The choice of technology is *hardly about "what works and what does not work"*. Purely going by traits, a number of technological options would have been possible to solve the pest problems of cotton. . . . Bt cotton's success belongs to the successful reproduction of the cotton-growing farmers' historically acquired and culturally consolidated ability to perform with the technology. . . . This *centrally challenges the notion of a smart, rational farmer* taking a correct decision in favour of his/her private and largely economic interests.

Similar to Latour, both Stone's and Shah's analyses show well that 'facts' do not simply 'exist', but come about in a process that is subject to a specific, to-be-decided-upon system of collection, demarcation, authentication and dissemination. Each system is 'structured by [its] own social conventions for creating forms of knowledge while nullifying others' (Stone, 2012: 63). However, the difference between Latour on the one hand and Stone and Shah on the other is that the latter, in typical anthropological fashion, are keen to explore the *social* and *cultural* systems that make 'facts' appear. Both are not interested in what constitutes the real 'truth' of what is claimed to be 'factual'. By contrast, as argued at length in Chapter 2, for Latour such an approach implicitly upholds the distinction between the house of nature and the house of the social – by analysing, in this case, social relations and dynamics without analysing at the same time 'natural' processes and their truth. As Latour argues in *Politics of Nature* (2004), if we truly want to leave behind the two-house logic, we need to make space for the 'perplexity' and 'surprise' that the 'natural' world can introduce into the social. In other words, Latour is not interested in how social conventions or particular historical conditions generate their *own* reality that can be contrasted to *other* realities. He wants to have *one* reality (constituted as the interaction of human and nonhuman) that is decided upon (through processes of admission and exclusion) in the collective. Science, for Latour, is not simply 'a method of inquiry' that should never be used as 'a political weapon or artefact of power' (Herring, 2006: 477). Instead, for Latour, science should definitely be 'weapon', insofar that it is part of the equipment used by those who formulate a new proposition to the collective.[1]

More specifically, in relation to the Bt cotton controversy Latour would not primarily critique the existing polarisation that leads to each side ignoring the conditions of its own truth construction. Instead, the problem is the reliance on 'facts' that both sides embrace, which closes down debate and disables different forms of speaking, listening and of making better, collective decisions. In this specific controversy Latour would ask for the 'voice' of the nonhumans that are involved: the cotton seeds, the water, the soil etc. On both sides of the debate, the agency given to nonhumans is a mere metaphorical one: seeds, for example, are supposed to 'feed', to 'sustain', to enable profit-making in contrast to the making of a living, to 'produce' and to 'reproduce' (see e.g. Shiva, Emani and Jafri, 1999; Shiva, 2009; Herring, 2006). But this representation is indeed metaphorical and not what Latour would call 'faithful': it simply serves to confirm existing narratives instead of challenging them.

Latour aims to give nonhumans a voice in the collective through making them part of associations that can formulate propositions (by, for example, giving them a 'speech prosthesis' – cf. Chapter 2). But as both post- and decolonial theorists have so aptly argued, it is far from clear that all *humans* have a voice. Is there something to be gained from Latour's understanding of a non-binary reality, of the voice of the nonhuman and of the speech prosthesis that allows us to have a different take on the *human* voice as well? And could this move us beyond the true/false binaries that this chapter has so far explored in relation to resistance to Bt cotton in India, beyond the mere observation that reality is socially constructed and relative?

Finding a voice in speaking through/with nonhumans

Who can speak? Revisiting Spivak's reflections on the subaltern

When Spivak revisits her famous essay 'Can the Subaltern Speak?' in 1999, she herself rejects the conclusion that many of her interpreters had drawn from reading the original piece in 1988, namely that the subaltern does indeed *not* speak. Spivak (1999: 255) explains that the real problem is the inscription of the subaltern as the 'oppressed subject' that can only speak for her/himself; an inscription that comes from the place of the intellectual who thinks s/he can 'abstain from representation' (ibid., 272). The voice of the subaltern is invoked without further reflection on how the intellectual's encounter with the subaltern (and her/his ability to listen to the subaltern's voice) is shaped by 'the relations of power involved in the colonial encounter' that provided the basis for the production of dominant knowledge (Bhambra, 2014: 128).

Learning to adequately represent *(vertreten)* the subaltern implies, for Spivak (1999: 276), 'to learn to represent *(darstellen)* ourselves', in the knowledge of our differentiated (hierarchical) geopolitical positions. If the subaltern is merely 'listened' to without an interrogation of the (historical, colonial) 'mechanics of [her/his] constitution', if 'authenticity' is invoked without an adequate analysis of structural constraints, we assimilate rather than understand the subaltern (Spivak, 1988: 294; cf. Bhambra, 2014: 128).

As is well known, Spivak develops this argument out of a close analysis of Foucault and Deleuze's famous conversation 'Intellectuals and Power' (1972). Her main critique of the two thinkers is their conscious rejection of the question of ideology, which implies a rejection of 'the difficult task of counterhegemonic ideological production' (Spivak, 1999: 255). Fetishising 'concrete experience' in a way that that leads to conflating 'one model of "concrete experience"' with '*the* model', Foucault and Deleuze, according to Spivak, unintentionally 'consolidate the international division of labor', because the oppression and domination resulting from the latter loses all specificity, becoming just another instantiation of the experience of power (ibid., 255–6; emphasis in original). Spivak's emphasis on the international division of labour is an important element of her revisiting the original essay in 1999; a year that is often taken to mark the start of anti-globalisation activism, manifested in the famous 'Battle of Seattle' in which thousands protested against the latest round of WTO negotiations. In the late 1990s, Spivak focuses on the link that she sees between historical imperialism and late twentieth-century processes of globalisation, which she calls 'New Empire'. For her, globalisation symbolises 'a return to the roots of imperialism's relation to capitalism where "free trade" agreements became the norm for conducting business' (Didur and Heffernan, 2003: 4; cf. Spivak, 1999: 311).

This diagnosis is not far away from Shiva's position and critique. Both Spivak and Shiva critique the domination of multi- or transnational corporations that contract out manual work to the 'developing' world, resulting in a world in which some people 'labour' but do not buy the products of their labour, while others 'consum[e]' (Spivak, 1999: 276). Despite these commonalities, previously mentioned problems with Shiva's approach still stand: the 'voice' that Shiva is willing to listen to when it comes to Indian farmers is completely pre-framed by her own analysis of the situation and her pre-given, unquestionable opinion on agricultural biotechnology. There is also little reflection in Shiva on differentiated geopolitical positions and how this constitutes spaces to speak. In her late 1990s work, Spivak defines 'the new subaltern' as being caught between either having to embrace global capitalism and development, or traditionalism and culturalism – with no space outside of this binary. This is a dilemma that Shiva is unable (and potentially unwilling) to address: for her the subaltern is inextricably bound to a (imagined) traditional culture unspoilt by colonial and capitalist domination in order to be authentic. There is no alternative other than global capitalism and the masculine logic of exploitation.

Spivak's critique of invoking the subaltern's voice as unproblematic is also interesting for an analysis of Herring's argument. As already pointed out, Herring and Rao (2012) bemoan the neglect of Indian farmers' voices in the Bt cotton debate without providing any reflection on how these voices are shaped by (historical, colonial) structures of power and domination. The voices of the farmers that Herring and his various co-authors listened to in interviews are indeed supposed, from Spivak's point of view, to render farmers transparent: if farmers say that all they care about is higher productivity, with technology being nothing but a potential tool, then this is all there is to know about the situation. Like Foucault and Deleuze (in Spivak's account), Herring depicts farmers as 'self-knowing, politically canny'; as subjects that can 'speak, act, and know *for themselves*' (Spivak, 1999: 257, 259; emphasis in original). While Spivak uses the example of the woman who 'know[s] that gender in development is best for her'; who knows that she needs to be ' "developed" . . . through UN Plans of Action' such as micro-credits (ibid., 259), Herring's farmers know that higher productivity is best for them and that science and technology are means of achieving that objective. For Spivak, Foucault and Deleuze's main problem lies in their failure to recognise how they are not able to 'merely report' on the self-representation and practices of the oppressed subject, but that their reporting takes place against an unquestionable analytical background that sets out to identify, in Foucault's case, 'the workings of . . . power', and in Deleuze's case the 'workings of . . . desire' (ibid., 265). Similarly, Herring's 'reporting' on the self-representation of the farmer takes place against an unquestioned background of agential rationality and a cost-benefit-calculation of interests.

What could or should be the consequences of Spivak's argument for taking a stand in relation to the Bt cotton controversy? We can find some indication in an Oxford Amnesty Lecture that Spivak gave in 2001, in which she critiques a tick-box approach to human rights education that aims to teach human rights to the subaltern without reflecting on the impact of 'centuries of oppression and neglect' on disenfranchised groups (Spivak, 2004: 548). An adequate reflection on the latter requires, Spivak (ibid., 549) argues, a 'painstaking foundational pedagogy which prepares the subject of rights from childhood and from within a disenfranchised culture of responsibility'; a humanities-based holistic education that can lead to what Spivak (ibid., 558) has termed an 'uncoercive rearrangement of desires'. This critique resembles the problems pointed out by some of the researchers who have engaged in ethnographic studies on smallholder producer practice in relation to GM crops. Matthew A. Schnurr and Sarah Mujabi-Mujuzi (2014), for example, suggest that too often voices of smallholders in Africa are assembled quickly, in straightforward interviews that address pre-framed ready-made questions that only allow for specific answers to be given (ibid., 644). Susana Carro-Ripalda and Marta Astier (2014) make a similar observation about research on smallholder maize producers in Mexico. It is to their impressive critique and reflections that this chapter will turn to next, in order to relate some of the critique on 'voices' that this section has made to the concrete context of agricultural biotechnology.

A question of ontology? Latour, Spivak and the nonhuman-subaltern voice

Schnurr and Mujabi-Mujuzi's point about voices being assembled too quickly by having to respond to pre-given questions relates well to Latour's critique of the conventional take on nature as revealing its truth through scientific means. Replacing the understanding of the scientist as detached objective observer of a natural truth that is already ready-made with the notion of the scientist as 'spokesperson', Latour defines reality as a dynamic conglomerate of humans and nonhumans that constantly form new interconnections. Humans can faithfully represent nonhumans via inserting them into scientific experiments that allow the latter to resist and turn around the questions they are asked by the former (cf. Chapter 2). Like Spivak, in *Politics of Nature* Latour does not altogether want to move away from representation, but interrogates the conditions under which successful representation can and must work. However, while Latour grounds it in the existence of a human-nonhuman-association that knows no subjects or objects, for Spivak there has to be a reflection on the *distinctiveness* of the position of involved subjects, which implies a reflection on existing relations of power and domination. Though these positions seem at a first glance mutually exclusive, they can be made compatible. For Latour there can be, by definition, no ontological distinction between the (human) subject and the (nonhuman) object – otherwise we would be back to the binary two-house logic. However, there is a *positional* difference between the human and the nonhuman in the context of the collective, which requires the former to act as a spokesperson for the latter in order to enable the formulation and deliberation of propositions. Bringing in this positional difference is not that far off from Spivak's emphasis on differentiated human positions, with the important qualification that the latter are grounded in concrete geopolitical history, while for Latour positional difference is (at least implicitly) grounded in differentiated morphological properties of humans and nonhumans. I argue that the concept of faithful spokespersonship can be taken up in the context of *both* historical geopolitical domination (leading to humans having to become spokespersons for other humans) *and* morphological differentiation (leading to humans becoming spokespersons for nonhumans). Faithful spokespersonship is defined by the extent to which both (subaltern) humans and nonhumans are able to disrupt, resist and transform the questions they are asked. This way, Latour's critique stops being merely an intra-modern one, because it becomes clear that the two-house logic is inextricably related to a colonial past that confined not just nonhumans, but (maybe more importantly) specific *humans* to a passive, objectified, to-be-discovered, -dissected and -exploited external reality.

All of this is significant for understanding the problem of 'voice' in controversies around the (lack of the) adoption of agricultural biotechnology by smallholder producers in the Global South. As Carro-Ripalda and Astier (2014: 659) point out, much of the research carried out in relation to these problems and questions is unable to grasp the 'ontological incompatibility' that exists between the experienced human/nonhuman relations in smallholder agriculture on the one hand, and the logic that underlies GE on the other – mainly because most research itself is bound by the latter logic. As Gregg Hetherington's (2013) reflections on his own anthropological research journey in Paraguay make clear, it can be difficult to understand the reality of the people you are interested in when you purely focus on the *social* dimension of statements that are made about nature. Coming from a position in which he took for granted the scientific distinction between (proven) 'fact' and 'error', and in which he also clearly demarcated his own objects of interests (social phenomena) from objects of interest to others (scientific facts), Hetherington explains how he 'translate[d]' claims of the leader of a local peasant movement (Antonio) about the truth of (GM) soy 'killer beans' into something else (ibid., 67; emphasis in original):

> Until this point, I had approached ethnography as an extended discussion with and about humans, and I was less interested in beans than I was with what Antonio said about them. . . . To be blunt, Antonio kept pointing at the beans, and I kept looking at *him* . . . I was comfortable saying that this was a figure of speech, a kind of political rhetoric, or even to claim that this is what Antonio believed, all of which explicitly framed 'la soja mata' (soy kills) as data for social analysis, rather than analysis itself worthy of response.

However, Hetherington (ibid.) points out that not believing in the truth of the killer bean did not prevent him from 'participating in Antonio's knowledge practices'. Hetherington became 'part of the situation' that finally pushed killer beans from being an errant knowledge practice to a matter of national concern, when two soy farmers were brought to court for the murder of two anti-GMO activist peasants. Crucially, participation involved more than joining the situation in spite of his lack of belief: participation conferred agency to the bean to such an extent that it finally managed to *speak* to Hetherington and thereby change the problematic, through the way it was encountered (ibid., 72):

> Beans didn't scare me at first. Indeed, as a foreigner to the situation that gives rise to killer beans (a Canadian no less), giant fields of soy were a familiar, even a comforting sight. But it took only a few months with Antonio for me to start feeling the menace from those fields. Soon, the sweetish smell of glyphosate, recently applied, and especially the corpselike smell of 2, 4-D mixed with Tordon, could ruin my appetite and make me expect to see people emerge from their homes to show me pustules on their legs and stomachs.

Similar observations are also found in Carro-Ripalda and Astier's contribution to the 2014 *Agriculture and Human Values* symposium on the challenges for smallholder farmer 'voices' in relation to agricultural biotechnology. While most of the contributions to the symposium concentrate on how to tease out farmers 'real' voices in the most effective way, based on the assumption that choices about GMOs are made by them as subjects, Carro-Ripalda and Astier critically reflect on their own perceived *failure* to get at farmers' voices in their research on GM maize cultivation in Mexico. Though these reflections, like Hetherington's, focus on agriculture and GMOs in a very different geographical, social, cultural and historical context, I maintain that useful lessons can be learned for researching smallholder producers' voices in relation to the Bt cotton controversy in India.[2]

It was through ethnographic fieldwork in rural areas in Central Mexico, in-depth structured interviews, focus groups, participant observation and, finally, a National Workshop in Mexico City with over 50 stakeholders (including small-holder producers) that Carro-Ripalda and Astier (2014: 656) attempted to get a better sense of what the actual voices of peasants in the GM controversy were trying to convey. However, particularly the final workshop, which aimed to cre-ate conditions under which Mexican smallholder producers could speak on their own terms about GM maize cultivation, 'unwittingly reproduced the conditions of exclusive, techno-scientific and regulatory spaces' (ibid., 660). This severely limited the way that smallholders could articulate 'their perceptions, ideas, and desires' (ibid., 659). The public discourse that centres on questions of safety, sci-ence, possibilities of regulation and problems of potential contamination, and that is upheld by *both* GM maize proponents *and* anti-GMO activists, dominated the workshop debate. This discourse relies on precisely the kind of two-house logic that Latour wants us to do away with: it upholds science as the most legitimate body of knowledge and limits a social/political discussion of science to its norma-tive obligations; it understands nature as separate, passive and external to humans; and it isolates 'the GMO' as a source of potential benefit or (on the anti-GMO side) harm. Inexpressible on those terms was, according to Carro-Ripalda and Astier (ibid.), the nature of the relation of farmers to their 'land, seed, crop, cli-mate . . . as told and understood by themselves'; the 'central place' that Maize continues to occupy in Mesoamerican pre-Hispanic cosmology, and 'the social and cultural significance' that goes along with that (Carro-Ripalda, Astier and Artía, 2015: 34). At stake for smallholders were 'their lives as maize cultivators, their pride in their craft and knowledge, and their ceremonially demanded right to information, choice and access to their "own resources"' (ibid.). It was not just about 'retaining "traditional" ways of agriculture', as the anti-GMO movement maintained, but also about claiming 'political, economic and socio-cultural rights' (Carro-Ripalda and Astier, 2014: 662).

Reading Carro-Ripalda and Astier's piece, it is occasionally difficult to under-stand what the problem of 'ontological incompatibility' really is about. After all, demands for rights, information, access and choice seem to chime quite well with conventional norms and frames. But this problem goes, I suggest, mainly back to the constraints posed by the format of the academic article itself, which requires a logically structured argument that culminates in a conclusion in which the main points (here the one about rights) are rehearsed and made memorable. Academic articles (at least in the social sciences) only tolerate so much dissonance, and can only convey sensual, non-linguistic experience to a limited degree. To prop-erly grasp the 'ontological incompatibility' that the authors describe, we need to focus on those passages that most strongly convey a sense of grappling; of not quite being able to put into words and argument 'the complexity of experiences, relations and reasons that bind people to maize' (ibid., 660). It is interesting, for example, that Carro-Ripalda and Astier talk about 'voices' as going beyond the semantic level, as conveying something acoustically, and as requiring a form of listening that shies away from asking pre-given question (ibid; emphasis added):

Despite the shortcomings of the workshop . . . we felt that that, through our research on the ground, we had engaged with male and female farmers, heard about their perspectives on GM and their visions of a rural future, and accompanied them to work in milpas and markets. So, *what do smallholder farmers' voices sound like?* What meanings did they convey to us? We will provide here but a few of those *sounds and meanings*.

The examples of actual 'voices' they then choose to present often involve speech prostheses given by farmers to the maize:

'It is a joy to plant, getting hold of the maize, of a beautiful cob which is pleasant, to go to the harvest, to look at pretty cobs, all regular. Because this is what sustains me.'

'You can see the difference in the seeds straight away. . . . You need to look at the cob and as soon as I grab it I see the difference.'

'It is the person who knows the seed the one who chooses it [for replanting the following year]'.

(all quoted in ibid., 660–1)

By contrast, GM maize is associated by the farmers whom Carro-Ripalda and Astier cite in their article with feelings of 'artificiality, estrangement and distrust towards the created object (the GMO) in itself, not only because of deep ontological considerations . . . but because of the political and economic motives which are "assembled" into it' (ibid., 661). The last point is important, because it shows that the particular political and economic situation that farmers face is part of the assemblage itself, and impacts on the sensual, bodily connection with the actual maize (as well as the other way round). The modified maize is a problem because it is part of a particular 'neoliberal vision of the Mexican agricultural future' (ibid., 658). However, that vision is not only (and not even primarily) made sense of intellectually, through given theories of e.g. the exploitation of labour, but sensually, through the way it disrupts the (physical) pleasure and joy that has sustained the farmer-maize-assemblage so far. GM technology externalises the maize from the farmer and alienates her/him from her/his way of life; and it is only through this externalisation that GM maize becomes perceivable as a potential source of 'contamination', as a danger against which the farmer needs to 'defend' her/his seeds (ibid.). The disruption also has an important temporal dimension, insofar that it does not only break the assemblage in the present, but also in the past and in the future, through destroying the connections of the farmer to her/his ancestors and to past as well as future generations (ibid., 661).

Now it could be countered that previous paragraphs only provide a fancy repackaging of the well-rehearsed arguments brought forward by anti-GMO activists such as Shiva: the fundamental problem of agricultural biotechnology is that it alienates farmers from their traditional, ancestral way of life, that it 'contaminates', that it allows for their exploitation and that it provides a further foothold for neoliberal visions of how the world should be ordered. There are two (related) points that I want to make in response to this potential allegation: first, providing a different ontology for making sense of phenomena is different to mere repackaging; even if the result of such analysis comes to similar conclusions. Secondly, there are qua definition limits to the way that a different ontology can be expressed in a language and an argumentative format that is inextricably bound to the dominant, to-be-contested ontology. Before I turn to the second point in greater depth in the next section, I want to elaborate on the first.

In my analysis, there is no pre-given understanding of neoliberalism and global power relations that dictates how the 'voice' of the farmer can be made sense of; and there is also no pre-given understanding of the GMO and the farmer. Indeed, the main problem of Shiva's analysis is its subordination to dominant ontology, which regards the alienating character of the GMO as an intrinsic property. Similarly, smallholder producers are regarded as intrinsic ' "reservoirs" of local or Indigenous knowledge or as "natural" conservators of biodiversity through their traditional practices', which 'unwittingly reinforce[s] images of smallholder producers as passive, timeless and voiceless' (ibid., 658). By contrast, the problem of 'alienation' in my analysis (following on from Carro-Ripalda and Astier's) has nothing to do with the (static) ontological properties of either GMO or farmer, but is immanent to a very particular historical-social-natural-economic assemblage that makes the GMO *become* an 'alien', dangerous entity. As Hetherington (2013: 70) argues, 'all of agriculture' is a

> kind of communicative waiting game among people, animals, machinery, plants, and weather. Growing crops is about interactions in which each action can be vaguely directed but is always already too implicated in complex processes to have predictable effects. . . . This is a view of agriculture and knowledge that is not burdened by ontologically stable objects or properties.

Grappling with 'ontological incompatibility' and its implications for the 'voice' of smallholders in Mexico, Carro-Ripalda and Astier have managed to become, I maintain, faithful spokespersons for both the subaltern and (by implication) the nonhuman: they have given both humans and nonhumans a voice that managed, as much as possible, to resist the pull of given questions and ready-made frameworks. Particularly through their reflection on their failure to make farmers 'speak' in their workshop, which they relate at least by implication to the dominance of particular ontological frameworks that are inextricably related to particular historical geopolitical systems of domination, they come close to the sort of representation for which Spivak asks. At the same time, pulling out the need for an altogether different ontology in order to enable farmers' voices to be heard brings Carro-Ripalda and Astier close to Latour's critique. In their analysis, the subaltern gains a voice through *speaking* (for the nonhuman) and *being spoken for* at the same time; and in so doing light is shed on existing (historical) relations of domination. The process in which this voice emerges is painstakingly slow, and it shies away from asking straightforward questions. It attempts to build up agency from below, in an open-ended way; with propositions being formulated by human-nonhuman-associations that cannot be judged within pre-given frameworks, but nevertheless have to take into account particular socioeconomic and historical structures.

Concrete 'solutions' to GMO controversies do therefore not lie in doing more research on the opinions of smallholder producers, as Herring argues. It is also not about advocating ready-made, fixed opinions on biotechnology and the nature of global problems, as Shiva does. Rather, it is about the continuous attempt to become faithful spokespersons for both nonhumans and subaltern humans – and often doing both at the same time – in the analysis of particular assemblages. Before I turn to such analysis in relation to the Bt cotton controversy in India, I want to come back to the problem of trying to pinpoint a different ontology within a format that is fundamentally grounded in the ontology that needs to be contested. I will do so by turning to Deleuze.

The wild being of statements and visibilities

Despite Spivak's well-placed critique of Deleuze's take on representation, there are affinities between Deleuze's thought and postcolonial studies, insofar that the former aims for liberation from the 'dogmatic image of thought' (Deleuze, 2004: 167) that characterises European modernity by creating new ontologies that transform the latter's points of confinement and exclusion (Chow, 2012: 157, 159). At the heart of Deleuze's philosophical project lies the aim to fundamentally overcome the universality of the subject-object distinction and the rigidity of identity, and to question the constant ordering of 'reality' along the lines of what is taken to be common sense. The dogmatic image of thought allocates clear positions for subjects and objects, understands the former as having to represent the latter, and marks out the Cartesian 'Cogito ergo sum' as what every self is supposed to understand intuitively (Deleuze, 2004: 164–213).

In his collaborative work with Félix Guattari, Deleuze argues that the role of philosophy is to create concepts that are not representable, insofar that they do no have 'spatiotemporal', but only 'intensive ordinates' (Deleuze and Guattari, 1994: 20–1). Understood in this way, philosophy itself can disrupt the dogmatic image of thought. The created concepts are incommensurable with any actual (political, social) phenomena; indeed, I would argue that the moment we use them to *represent* actual phenomena we are in danger of slipping back into the dogmatic image of thought. Taking up Deleuze and Guattari in this way is unfortunately quite common: IR theorists (e.g. Doty, 1999) use the concepts of nomadic life and the apparatus of the State in order to talk about *literal* sovereign states and *literal* movements of people, and many social movement scholars enthusiastically emphasise the 'rhizomatic' character of movements and events such as Occupy Wall Street or the World Social Forum (see e.g. Funke, 2012; Tormey, 2012). For Deleuze, by contrast, concepts cannot represent anything – neither literal migrants nor protest movements. All they can do is 'to direct [. . .] [our] thoughts – as if they were hands – over the flow of the virtual so [they] can swoop down and pull out this or that thing into a newly formed assemblage' (Strathausen, 2010).

For Deleuze, concepts are 'things' themselves: 'things in their free and wild state' (Deleuze, 2004: xix); objects that do not represent, but encounter. The task of thinking is to *create*, to carve out (new) assemblages, instead of reflecting on those that are already there.

Understanding the task of philosophy in this way is in need of a particular metaphysics. Traditionally, metaphysical analysis establishes a coherent, fixed system of the fundamental nature of the world that enables the latter's legislation from the outside. However, this view would be at odds with Deleuze's intention of questioning universalism and rigidity. Consequently, Deleuze, according to James Williams (2005: 4–5), faces the task of developing a 'metaphysics as a dynamic structure', which is achieved by understanding reality as two different fields of reality: the 'virtual' and the 'actual' (see e.g. Deleuze, 2004). The virtual field is expressed in the actual via a process of actualisation that is not entirely contingent, but conditioned by what goes in the virtual. Although virtual and actual fields are connected, they are different *in kind*, which means that the laws and processes according to which different elements relate in the virtual field are not the same as the ones in the actual. This implies that the former does not simply translate into the latter, which means that both cannot be compared on an equal-level playing field, and that the virtual cannot be privileged *over* the actual. Instead of setting up a 'new' system in opposition to the 'old', Deleuze aims to envelope the old in the new; which implies that not even Deleuzean philosophy itself can be conceptualised as that which is *different to* something else. Becoming has been 'render[ed] . . . ontologically independent from being' (Strathausen, 2010).

The problem with Deleuze's critique of modern rationality and his metaphysics is, as already pointed out in Chapter 1, that it remains intra-modern, insofar that it neither takes into account how modernity and coloniality are co-constituted, nor that overcoming it requires a move beyond modern territory that is not just philosophical, but that needs to engage with the thought (and ontologies, and metaphysics) of non-modern traditions that modernity has ignored and oppressed (cf. Todd, 2016; Mignolo, 2007). However, I argue that in relation to the problem of accessing ontological difference that the previous section has pointed at, Deleuze's metaphysics can be helpful insofar that it gives us a wider, beyond-the-actual understanding of reality that allows us to *sense* and follow what Deleuze and Guattari (2004) call the line of flight; forming new connections *in* and *through* the creation of new concepts and assemblages that are able to come to terms with ontological difference at a non-categorical and more-than-linguistic level. Without such metaphysics it becomes difficult to conceive how we can even *access* thought and understandings from the 'outside' of modernity without straightaway subordinating them to the dogmatic, modern image of thought (cf. Viveiros de Castro, 2015).

For the purpose of this chapter, I want to engage with Deleuze's book *Foucault* (1999), in which Deleuze himself deals with the question of 'voice', as well as Deleuze and Guatarri's plateau 'On several regimes of signs' in *A Thousand Plateaus* (2004), which engages with the role of language. Analysing Foucault's understanding of discourse in the latter's archaeological work, Deleuze's *Foucault* attempts to make sense of the relation of the discursive and the non-discursive in the construction of what Deleuze calls historical strata. The latter are hierarchical structures organised by principles of identity and representation that substantialise the formless matter of the virtual. When engaging Deleuze's *Foucault*, previously outlined metaphysics needs to be kept in mind: Deleuze does not engage with other authors in order to give yet another, 'truthful' interpretation of their work, and to develop his own work based on objections or agreements with them. Precisely because philosophy is, for him, all about the creation of concepts of difference in a non-binary way, he takes what others offer as creative starting-points for the development of new concepts. In *Foucault*, Deleuze uses Foucault's work by subverting it, using it for his own purposes and supplying it with new concepts and ontologies, without ever making that move explicit. One striking example is the swiftness with which Deleuze takes the empirical (historical) element out of Foucault (which is so crucial for Foucault himself) and understands Foucault's archaeologies as *creations*: 'History responds only because Foucault has managed to invent . . . a properly philosophical form of interrogation which is itself new and which revives History' (Deleuze, 1999: 42; note that the capitalisation of 'History' seems to be a deliberate move).

In *Foucault* Deleuze points out that a historical stratum is made from 'things and words, from seeing and speaking, from the visible and the sayable' (ibid., 41). However, in order to get at 'the statements' and 'the visibilities' of the stratum, actual words, phrases and perceptible 'things' must be broken open in order to reveal what is not 'directly readable', 'sayable' or 'visible' (ibid., 46). Statements and visibilities must be *extracted* from actual words and perceptions, though Deleuze is at pains to emphasise that this does not mean that the former have been *hidden* (ibid.). In contrast to words and perceptions, statements and visibilities are about the 'there is' of language and light; the 'light-being' and 'language-being . . . which is not to be confused with any of the directions to which language refers', or any object that 'would show up under light' (ibid., 45, 48, 50). In relation to statements, Deleuze points out (based on a reading of Foucault's *Archaeology of Knowledge*) that they only 'become readable or sayable . . . in relation to the conditions which make them so and which constitute their single inscription on an "enunciate base"' (ibid., 47).

A Thousand Plateaus enables us to get a better sense of what the 'language-being' or 'light-being' of the statement and visibility might be about. Here, Deleuze and Guattari (2004) argue that that the directly sayable (the proposition) is structured by various regimes of signs, which in turn are related to 'abstract machines, diagrammatic functions, and machinic assemblages that go beyond any system of semiology, linguistics, or logic' (ibid., 163). Abstract machines and diagrams are concepts that are similar to what Deleuze calls the virtual in other parts of his work: they refer to unformed matter that does not distinguish between a 'thing' (or substance) and an expression *of* that thing: 'it is no longer even possible to tell whether [something] is a particle or a sign' (ibid., 156). In other words, everything simply *is*, or more accurately, *becomes*. Abstract machines are, one could argue, fully virtual. By contrast, machinic assemblages have 'two poles or vectors': one 'is oriented towards the strata' and the other towards 'destratification', towards a 'deterritorialization' that is carried 'to the absolute of the earth' (ibid., 158, 160). Out of this understanding emerges, for Deleuze and Guattari, a three-fold task: the first is to extract from propositions the various regimes of signs that make the proposition possible. The second task lies in opening up to the abstract machines/diagrams that lie 'behind' the regimes of signs (ibid., 163–4), and to follow the line of flight towards absolute destratification. The third (simultaneous) task lies in following the second vector of machinic assemblages – the one oriented towards strata – to create new strata and statements that in themselves move beyond what is 'sayable' right now. At the end of this process we might indeed be closer to what Deleuze, in his book *Foucault*, calls 'language-being' and 'light-being': we might now have 'a patois of sensual delight, physical and semiotic systems in shreds, asubjective effects, signs without significance where syntax, semantic and logics are in collapse'; covering regimes of signs as well as 'cries-whispers, feverish improvisations, becomings-animal, becomings-molecular' (ibid., 163). The second move remains irreducible to the third; both are 'inseparable', entertaining 'perpetual relations of transformation, conversion, jumping, falling, and rising' (ibid.).[3]

Some of these moves are detectable, I argue, in Carro-Ripalda and Astier's reflections on ontological incompatibility – particularly in their frustration about the limits imposed on the 'sayable' at their stakeholders' workshop. It can also be found in Carro-Ripalda and Astier's sensibility for the becoming that lies 'behind' the words and propositions that they have at hand. From the perspective that I have outlined here, Carro-Ripalda and Astier indeed perform the first move in the process of creating new statements: they sense that there are different regimes of signs that structure the concrete propositions that the farmers make. One of these regimes resembles what Deleuze and Guattari call the 'presignifying' regime of signs, which 'fosters a pluralism or polyvocality of forms of expression' (ibid., 130). The presignifying regime of sign entails 'forms of corporeality, gesturality, rhythm, dance, and rite' that 'coexist heterogeneously with the vocal form' (ibid.).[4] Sensing such a regime of signs in the propositions that they hear, through walking, listening and participating, might be precisely what leads Carro-Ripalda and Astier to commenting on the challenge of ontological incompatibility.

To take another example: in Hetherington's narrative of the situation of farmers in Paraguay, he (2013: 69) mentions a cotton farmer – Ortega – who said in a conversation that his success in saving his yield from pests 'was about being aware of the needs of his crops at all times' (ibid.; emphasis in original):

> Late at night, Ortega told me, he could hear his cotton crying *(hase)*, and he was unable to sleep. The cry demanded a response, but he lay there helplessly waiting for the sun to come up so that he could strap on his pesticide pack and go tend to the problem. . . . The ability of both Ortega and his cotton plants to respond to each other (their "response-ability" [referring to Donna Haraway]) may decide whether either of them is here next year.

The proposition 'I am aware of the needs of my crops at all times' is first of all structured by the regime of signs that Deleuze and Guattari describe as the despotic, imperial, signifying one. The latter is characterised by chains of signifiers that are pulled towards a despotic centre that the signifiers are supposed to recharge through multiplication, without ever referring to a 'real' signified (there is only an 'amorphous continuum that for the moment plays the role of the "signified," but it continually glides beneath the signifier' (Deleuze and Guattari, 2004: 124)). Talking about the 'needs' of the plant does not refer to a nonhuman agent 'plant' that expresses 'needs', but is part of a signifying regime in which farmers as subjects tend to plants as passive objects and thereby stabilise the dogmatic image of thought. On the other hand, the proposition is also structured, as the longer quote makes clear, by a different regime of signs in which, similarly to what we can find in Carro-Ripalda and Astier's account, vocal expressions coexist heterogeneously with other forms (a response-ability between farmer and plant that goes beyond the vocal, that is corporeal and takes place at the level of 'sense'). Detecting other regimes of signs creates space for transformation, and for a moving closer to the abstract machines and machinic assemblages that lie 'behind', leading to the creation of new statements: regimes of signs but also becomings-cotton, becomings-pesticide, whispers and cries.

Here, the disruption of the 'modern' dogmatic image of thought is not related to a particular way of doing agriculture (traditional versus modern, organic versus conventional). Indeed, the solution that the farmer finds to the problem of pests (namely the use of pesticides) could be depicted as straightforwardly modern, in line with assumptions that the likes of Herring hold about farmers' agential rationality and empirical pragmatism. Instead, disruption occurs in and through the space that is created for a different understanding of reality, and for encountering what Carro-Ripalda and Astier call 'ontological incompatibility' in the process (which, for Deleuze, is not so much about incompatibility but rather about a wider understanding of what 'real' multiplicity is about).

The danger of linking Deleuze's metaphysics to practical examples and experiences in relation to questions of agricultural biotechnology is that the former easily becomes yet another framework of representation for the latter, which harbours the danger of epistemic (and in the context of this book, colonial) violence. In addition, in relation to the examples that I have used here, it could be alleged that I have once again focused on the experiences of Western-based researchers encountering difference, detecting other regimes of signs, extracting and creating new concepts. However, the latter allegation takes for granted a *modern* understanding of agency, insofar that we understand Carro-Ripalda, Astier and Hetherington as *initiating* the transformative process described here. Instead, from Deleuze's perspectives (which is one that would be affirmed by Latour), all of them merely participate in human-nonhuman assemblages that distribute and diffuse agency. Indeed, the argument that we need to follow the 'voices' and 'practices' of *human* non-modern subjects in order to disrupt modern/colonial thought and practice in itself relies on a modern understanding of history, agency and subjectivity. Moreover, in this chapter the merits of Deleuze's metaphysics are not judged on the basis of how successfully it disrupts modern rationalities in the abstract, but to what extent it can contribute to making space for different (or even incompatible) ontologies, regimes of signs and statements that have been ignored, excluded and annihilated in the process of modernity establishing itself as universal. In other words, Deleuze's metaphysics is relevant only if it does not just question the dogmatic image of thought as a *modern* one, but also as a *colonial* one; the possibility of which this section has shown.

States and machines: thinking differently about Bt cotton

It is striking that many reflections on ontological incompatibility in relation to smallholder producers and agricultural biotechnology in the Global South come out of research conducted in Latin America. Indeed, I have been unable to find similar reflections on the Bt cotton controversy in India. The primary reason for this might lie in the particular research interests and frames of those who have conducted fieldwork in India.[5] But there might also be another reason for the silence of Indian smallholder producers in relation to Bt cotton.[6] As I have already indicated, it is important to pay close attention to the specificities of India's colonial legacy, its postcolonial history in relation to state-development and economic structures, and its cultural and social context. What I want to argue in this section is that in relation to agriculture, the specific context that we have in the case of India has led to reducing the space for the possibility of ontological incompatibility – a reduction that has already started during colonial rule, has then been advanced after independence with the building of an (initially) socialist nation-state that had fully incorporated Western notions of progress, development and science, and has been reinforced by the opening up of India to the market forces of economic liberalisation.

According to Sudipta Kaviraj (2011: 36), India's assessment of European modernity and the latter's 'primary historical instrument' – the state – was impacted on by two different, opposing strands of thought: Bhudev Mukhopadyay's essays on sociology and Mathatma Ghandi's theories. Kaviraj (ibid., 37) points out that despite his general rejection of European statehood, Mukhopadyay confirmed and admired two of its features, namely political economy as the pursuit of improving a nation's wealth, and the support of modern science. The combination of both has had a strong impact on the development of Indian statehood after independence, when the belief in economy and science was coupled to the production of a new powerful imaginary about the capacities of a successful postcolonial state. This resulted in the development of a 'vast bureaucratic regulating machine' (ibid., 39; Sinha, 2008: 73). As K. Sivaramakrishnan and Arun Agrawal (2003: 37) point out, 'the nation-state was always a linchpin of development in the first phase', and even in the second phase (characterised by the shift to market forces) 'it seems scarcely credible that the nation-state, its agencies and personnel, and other national level actors, have ceased to be significant players' (cf. Rosenow, 2013).

With the possibilities offered by the Green Revolution in the 1960s and 1970s India shifted its approach towards agriculture from a position in which it emphasised the need to overcome the reliance of the economy on agriculture towards one that embraced it as a central means to combat poverty (Sinha, 2003: 294).

Unlike Mexico, where agriculture accounted for 9 percent of GDP in 1980 and 4 percent in 2010, 36 percent of Indian GDP came from agriculture in 1980 and 19 percent in 2010. Until today, 'India's population remains predominantly agrarian and rural', with 55 percent of the total workforce having worked in agriculture in 2011 (by comparison, only 22 percent did so in Mexico) (Egorova, Raina and Mantuong, 2015: 105). The state's emphasis on scientific progress as solving the problems of agriculture and therefore also of poverty has become deeply engrained in the self-understanding of Indian farmers; particularly those whose wealth has been built on the success of the Green Revolution (Shah, 2008; cf. Sinha, 2003: 294). As Shah points out in his study of resource-rich farmers in Gujarat, anxieties that always accompany agricultural practice – the finiteness of natural resources such as land and water and the unpredictability of nature in the form of weather and pests – are often 'compensated by unlimited faith in science and the state' (Shah, 2008: 442). When Shah (ibid.) attempted to challenge these farmers in that belief, by pointing out that pests would always develop resistance to the newest engineered cotton variety in the end, his scepticism was countered by statements such as '[s]omething else will emerge'; 'scientists would come up with some more research and advanced methods'; and 'if scientists don't know certain things now, they will soon know' (ibid.).

With regard to the planting of cash crops such as Bt cotton, the possibility of a different 'relation to the land', a different 'way of life' such as the one that Carro-Ripalda and Astier describe in relation to smallholder maize farmers in Mexico, which grounds much of the scepticism and resistance towards biotechnology that we can find there, seems diminished.[7] As researchers of the *GM Futuros* project have observed in their comparison of Mexico, Brazil and India, Indian actors (including NGOs and smallholder farmers) who were opposed to biotechnology strongly grounded, in contrast to biotech opponents in the other countries, their arguments in science and biosafety concerns that 'only poorly and imperfectly capture the social, ethical and political stakes of the issue' (Egorova, Raina and Mantuong, 2015: 113; cf. ibid., 119). Even Shiva, when interviewed by the *GM Futuros* researchers, mainly drew on science for making her argument (ibid., 120). Interestingly, the apparent lack of cultural sensitivity changed when it came to the possibility of GE food (ibid., 128). Indeed, the 2013 Indian moratorium on agricultural biotechnology has been the result of controversies and regulatory problems around the attempt to commercially release GE aubergines (Bt brinjal) (ibid., 67) (though note that field trials of GM food crops have re-started in 2014; Egorova, Raina and Mantuong, 2015: 109). The aubergine is a crop of high symbolic value for Indian national identity (Egorova and Mantuong, 2014: 76). That the issue might be a different one for Indian farmers when it comes to food shows that a different 'relation to land' might (still) be present in India when it comes to particular plants (after all, the issue in Mexico is also about a subsistence product – maize). But that many farmers and peasants in India rely on the planting of cash crops such as cotton instead of subsistence products also tells us something about how agriculture in India has developed since the advent of the Green Revolution.

The strong belief in the potential of science and technology is also grounded in Indian agricultural social relations and economic structures since colonial times. As Shah (2008: 438) points out, in the state of Gujarat that he studied in detail, access to land had been determined under colonial rule by the land-allocating British elite. They privileged the caste of the Patels; a privileging that still grounds the ongoing economic and social dominance of that caste today. Similarly access to water, which determines the success of growing cotton (against common perceptions cotton is usually irrigated), has been historically determined by the British ground water extraction policy, which 'was so designed that only wealthier cultivators could afford to dig a well in the first place, and then pay the exorbitant taxes levied on it' (ibid., 439). Another colonial legacy has been the introduction of the American variety of cotton, which was used because it was more suited to European machinery and therefore desirable for the British, though by contrast to the native *Desi* variety it 'was highly susceptible to pest attack' (ibid., 437). The advent of the Green Revolution then made the native variety completely uneconomical, because the Green Revolution relied on the use of fertilisers to which the native variety was unresponsive (ibid.). Also, those farmers who have profited in the past and continue to profit today have had 'access to labour surpluses': migrant labour that has become available during the intensification of agriculture in the 1970s, and that today is largely made up of young women and children (ibid., 439).

Shah's analysis is confined to Gujarat, and there are obviously differences to other states and regions in India (ibid., 444). In the Vidarbha region, for example, disappointing results in relation to Bt cotton might have been due to 'an intense lack . . . of the cultural and social solidarity found amongst Gujarat's elite cotton-growing farmers' (ibid.). As discussed earlier in this chapter, Warangal in Andrah Pradesh has been characterised by a lack of social and environmental learning resulting in 'deskilling' (Stone, 2007; cf. Shah, 2008: 444). Despite regional differences, the thrust of Shah's observations chimes with my general elaboration on the development of the Indian state after independence. While among the Mexican smallholders the prospect of GE maize was perceived to be an outside force that *disrupted* and *cut off* the way farmers connect to their land and themselves, the situation in India could be described as one of *continuity*. Indeed, the need to not having to adapt to a different technological-cultural paradigm (due to the way biotechnology seamlessly links to the dictum of the Green Revolution) is raised by Shah as one of the major reasons for the adoption of Bt cotton in Gujarat.

The State, Man and the ethereal gene

Coming back to Deleuze (and Guattari), if we want to turn the previously identified colonial continuity that we find in the Indian approach to agriculture into a disruption, we need to once again extract the various regimes of signs that we find in the propositions that stabilise that continuity. In the previous section of this chapter I have already referred to two regimes of signs that Deleuze and Guattari elaborate on in *A Thousand Plateaus*, and I have extracted those regimes of signs from the propositions that I have found in Carro-Ripalda, Astier and Hetherington: the signifying, despotic and imperial one (1), and the presignifying one (2). Altogether Deleuze and Guattari refer to the existence of four regimes, and for the purpose of this section it is helpful to also explain the remaining two. As I have already mentioned, Deleuze and Guattari understand all propositions as being structured by a mix of regimes. Indeed, as they make clear towards the end of the plateau 'On various regimes of signs', the first step towards the opening up of propositions and the creation of new statements that incorporate becomings already lies in extracting not just one, but several regimes of signs from each proposition (Deleuze and Guattari, 2004: 162). Despite all regimes existing in a mix, a particular regime can be dominant in relation to specific (historical) strata. Deleuze and Guattari (2004: 149) make clear that for what they call the apparatus of the State, the dominant regime of signs is the previously-explained signifying, despotic, imperial one. However, it is in relation to the apparatus of the State that they also identify a third regime of signs: the 'countersignifying one' (3), which is the semiotic of the 'war machine' directed against the State (ibid., 131).

Similarly to the relationship of the virtual and the actual, the war machine and the State apparatus are not simply oppositional, but different in kind: 'In every respect, the war machine is of another species, another nature, another origin than the State apparatus' (ibid., 389). Like the 'abstract machine' more generally, the war machine is asubjective, made of 'material-forces' rather than 'matter-form'. It is destratified, occupying smooth space (ibid., 407). The war machine is not simply external to the State, but in itself 'a pure form of exteriority' (ibid., 390). This means that we cannot take the State as a 'model' against which the war machine becomes conceivable as external and 'negative'. Instead, we have to leave behind the very categories of the State in order to make sense of the war machine (ibid., 390–1). This is an argument that is in many ways similar to the decolonial argument about the need to de-link from modernity. For Deleuze and Guattari, the State apparatus is not primarily an actual historical form, but a principle that is manifest in the dogmatic image of thought: it is a particularly rigid way to segment life, enabling order and domination.

Focusing on the *spatiogeographic* as one aspect that makes the State different in kind from the war machine, Deleuze and Guattari (ibid., 420) contrast the latter's smooth 'nomadic trajectory' with the striated 'sedentary road' that defines the State; the function of which is to create a 'closed space', in which, as the dogmatic image of thought dictates, each person is allocated a share, and in which the communication between shares is regulated by external rules. Striated space is metric; it 'plot[s] out a closed space for linear and solid things' in entirely Euclidean fashion. By contrast, smooth space is vectorial, projective or topological: it occupies without the possibility to count, it is open and distributes 'things-flows' rather than entities (ibid., 399). Smooth space 'holds space and simultaneously affects all of its points', rather than counting the points and determining movements as that which takes place between the points (ibid., 401).

With regard to the second, *arithmetic* aspect, Deleuze and Guattari (ibid., 430) point out that the State uses the number to enable the definition and control of calculable objects in space and time, submitting variation and movement to the State's spatiotemporal (territorial) framework. By contrast, for the war machine the number does no longer refer to objects outside of itself, but 'marks a mobile and plural distribution': it arrives at 'arrangements rather than totals' (ibid., 131). The number becomes a central mechanism in the determination of smooth space. The third element that Deleuze and Guattari describe as characteristic for the State apparatus is the arranging of space by the State in a way that brings opposing forces to a state of equilibrium, while the principle of affection that reigns in smooth space does prevent such state (ibid., 438).

For those who have been supportive of the Green Revolution (in India and beyond), the space of nature is striatable rather than smooth, in line with the logic of the apparatus of the State. It is closed and characterised by clearly demarcated figures that follow ever-lasting upward spirals of development and depletion facilitated by technological invention. Movement takes place from point to point and can be calculated: in relation to Bt cotton, every hybrid series is followed by the next; each coming with new pesticides, and each 'slack[ing] after cultivation for 5–7 years' – '[a]nd so it goes on' as one of Shah's informant farmers optimistically concludes (Shah, 2008: 438). A new cotton variety is followed by a new set of cotton worms, which is yet again followed by a new cotton variety: a 'continuous interplay between the artefacts – new cotton varieties and pesticides – and nature's agency – worms' (ibid., 438). Cotton, pesticides and worms are understood as fixed entities with inherent properties that interact according to a set of external universal laws of equilibrium; giving, through such striation, stability to a system of production that is subject to ultimately unpredictable forces.

Deleuze and Guattari's concept of the State apparatus can be linked to what I have tackled in Chapter 1 as the 'survival machine' proposition that, in its emphasis on 'gene centrism', sustains a Western world view that emphasises 'the persistence of the eternal soul, or order and stability in the face of change' (Oyama, 2000: 1). As I have pointed out in Chapter 1, based on the witness testimony of Sylvia Wynter, 'survival machine' is a proposition of overrepresentated Man – Man/gene – the identity of which is/was based on a history of colonial exploitation and eradication of the 'Human Other.' The State-related stratification of nature that I have described here, and that is precisely what underlies the Green Revolution and the development of the Indian state before and since independence, is working along the same lines. As Shah (2008) points out, among the resource-rich farmers in Gujarat the GM Bt parent plant, which (ironically enough) was indeed the *male* parent (to be cross-pollinated with a female hybrid; ibid., 436), was 'put on a pedestal of immortality': 'Bt male was claimed to be not ageing, ethereal. These farmers believed that if allowed to self-pollinate without any contamination, Bt male's genetic capacity would last forever' (ibid., 442). The resemblance of this belief with the Central Dogma understanding of the gene as central, unchanging agent, and the concept of overrepresented colonial Man is striking. Indeed, the time of the Green Revolution in which this belief took hold was the same time in which India became a nuclear 'superpower', which created a social imaginary that provided the nation 'with masculinity, respect and power' (Visvanathan and Parmar, 2003: 2714).

The silence of the subaltern and the noise of the circus

All of the above is the background against which we have to assess and understand the (epistemic, but also often literal) silence of not only Indian smallholder farmers, but (maybe more importantly) of the landless migrant men, women and children on the fields – a silence that is sustained through particular social and economic power relations, but also through the deeply engrained colonial Man/ gene/State apparatus the roots of which have been planted during colonial times.

I have argued in the first chapter of this book that a decisive problem of some so-called New Materialist approaches is the failure to be able to integrate an idea of 'lack', of 'without', and of differentiation into their ontologies of connectivity and abundance. This chapter has pointed out that the Man/gene/State apparatus and the way it penetrates the reality of being in India has precisely led to lack: a *lack* of ontological alternative, a *lack* of a voice that is able to go beyond or outside of that apparatus and related regimes of signs. Too much emphasis on how to give voice, how to (re-)connect and assemble anew – in other words, too much focus on how there are *always* statements, machines and assemblages to be extracted from words, and visibilities from perceptions – is in danger of too quickly bridging that gap and filling the lack. Too much emphasis on the universe as a 'vibrant world of colour and form, of light and music' (Ho, 1998: 76) drowns out the spaces that are deafeningly silent. By contrast, what the situation in India in relation to the Bt cotton controversy requires might be an enduring of the discomfort of silence. This is similar to what Spivak requests us to do in relation to the education of the subaltern, which is a work that is painfully slow, not oriented towards outcomes and, as Spivak emphasises, needs to take place outside of the noise that usually accompanies her own theoretical work (she mentions that she usually does not bring her work with the poorest in India into the public realm at all). Maybe paradoxically, the concept of faithful spokespersonship might precisely enable us to *not* fill the silence too quickly: as I have already outlined in several places in this book, for Latour faithful spokespersonship requires the to-be-represented nonhuman (or, as I have pointed out in this chapter, the to-be-represented human subaltern) to be able to resist and transform the questions it/s/he is asked. As I have argued in Chapter 2, the problem of the social sciences is that they find it hard to endure silence in response to the questions that they ask. Too quickly do they define away the resistance of the material that they engage. From that perspective, a faithful spokesperson for the subalterns in relation to the Bt cotton controversy might not be the one doing new field research in India, but precisely the one who has nothing to add to the silence that s/he has encountered, a silence that brings out the fundamental nature of the colonial violation.

Coming back to the anti-GMO resistance with which I have started this chapter, it could be argued that the level of 'noise' of the former is precisely in danger of drowning out that uncomfortable, necessary silence. Indeed, the controversy around agricultural biotechnology could be considered as one of the noisiest of controversies that there is/was in India. As Shiv Visvanathan and Chandrika Parmar (2003: 2715) point out, it is 'a great morality play, a socio-drama of positions, a circus of spectacle'; it is 'folklore', 'gossip', 'rumour'; 'an orchestra of positions', a 'soap-opera'. Although there is an ongoing focus on scientific arguments, anti-GMO peasant organisations and grassroot NGOs very well make strong connections to neoliberal economic policies, the withdrawal of the state and state regulation working in favour of corporate interests, as the arguments of Shiva and the KRRS that I have outlined in the beginning of this chapter make clear. Is there really no resistant force at all in the noise of that circus? Taking into account Deleuze's argument in *Foucault* that the concept of the statement is 'inspired by music' to the same extent that the concept of visibility is inspired by the 'pictorial' (Deleuze, 1999: 45), the question emerges whether the sound, murmur, music and colours of Indian anti-GMO acts of resistance, which have managed to connect 'university debate, street theatre, religious discourse and a policy document' (Visvanathan and Parmar, 2003: 2715) are not able to fundamentally disrupt State stratifications and the problem of continuity after all. Does it fill too quickly the silence created by the State/Man/gene apparatus? Or are the assemblages that we find here resistant after all?

In his 1996 essay 'Anxious Hindu and Angry Farmer: Notes on the Culture and Politics of Two Responses to Globalization in India', cultural critic D. R. Nagaraj (2012) explains the protests of the Karnataka State Farmers Organisation (KRSS) against a Kentucky Fried Chicken (KFC) branch in Bangalore as having been based on what he calls the image of 'hunger-based identity'. This identity, Nagaraj (ibid., 287) maintains, refers to '[t]he creation of hunger' as being 'one of the foundational states of globalization' and the 'violence' that it 'causes in its modes of production'. As Nagaraj points out, the mode of the protest can only be understood against the background of the traditional democratic workings of India's polity, which up to the mid-1980s had been based on 'the active and effective presence of three interlinked symbols: the prison, the spade, and the ballot box' (ibid., 301; note that Nagaraj draws on Rammanohar Lohia's theory of democracy here). Up to that time, the best way to gain access to institutionalised legislative processes – and in the case of the KFC protests the elections to the state assembly were just 'round the corner' (ibid., 300) – was, indeed, through street protests and prisons (into which the leaders of the anti-KFC protests duly went): 'there was no significant difference, in terms of political time or space, between streets bursting with political demonstrators, overcrowded prisons, and a simple majority at least in the legislature in parliament' (ibid., 301). The immediacy of the connections between political protest ('the prison'), legislative politics ('the ballot box') and meaningful community action ('the spade') was connected to the immediacy of everyday objects (the KFC chicken) and high politics. Gandhi, Nagaraj (ibid., 301) points out, used the everyday object of salt, and with that 'shook the foundations of the British empire.'

It is in this tradition that we also have to understand the 'circus of spectacle' of the protests against biotechnology, and it is in this tradition that we have a potentially *productive* colonial legacy. As Ranabbiar Samaddar (2010) argues, 'ruthless colonial rule moved the colonised societies to a resistance culture where the normal question to be asked would be: Who are you to rule?' (ibid., xii), which grounds the connection between the 'prison' and the 'ballot box' as well as an understanding of politics that necessarily needs to be studied as 'material forms': 'the physical nature of rule', the 'material of organs of governing', the 'violent genesis of laws', and the 'contentions through which all politics proceed' (Samaddar, 2007: 3). Linking this to Nagaraj's reflections and to Deleuze and Guattari, I argue that the materiality of politics and the way it connects and assembles various spaces, forms, sounds and colours form a war machine that is (in kind) heterogeneous to the striations of the State. A space that is designed to be closed – the prison – becomes a smooth space; forming a machine with other elements that is, as Claire Colebrook (2002: 56) argues in relation to Deleuze's notion of the machine, 'nothing more than its connections; it is not made by anything, is not for anything, and has no closed identity'. The protesters-prison-spade-ballot-box-war-machine distributes humans and nonhumans in an 'open', indefinite space, which is not characterised by *movement*, which is part of a body that moves from point to point as 'one', but *speed*, which

> constitutes the absolute character of a body whose irreducible parts (atoms) occupy or fill a smooth space in the manner of a vortex, with the possibility of springing up at any point.
>
> (Deleuze and Guattari, 2004: 420–1)

While striated space defines bodies according to their weight when they fall, a vortex that occupies smooth space defies gravity at least to a certain extent: it is characterised by the deviation of the body 'from its line of descent or gravity' (ibid., 409–10).

However, for Nagaraj, the KRSS protest against KFC has failed to take into account how this traditional way of doing politics in India has been undermined by the process of globalisation, which, as capitalist axiomatic (Deleuze and Guattari, 2004: 486), has disrupted and then reconnected the familiarity of everyday objects such as chicken. Following Deleuze and Guattari's narrative (ibid., 486), the State apparatus operates via *overcoding*, while capitalism, by contrast, is an *axiomatic* that functions via *decoding*. From the very beginning, they argue (ibid., 499–501), capitalism has been able to mobilise 'a force of deterritorialization' that 'infinitely' went beyond the deterritorialisation the State itself could release. Capitalism became a world axiomatic of decoded flows, because private property and capital gained the independence of a Subject, independent from the State (ibid.). While 'all manner of codes, overcodings, and recodings' are relative to domains (e.g. territory as the decisive domain of the State), the axiomatic 'deals directly with purely functional elements and relations whose nature is not specified, and which are immediately realized in highly varied domains simultaneously' (ibid., 501; cf. Rosenow, 2013). In the despotic, imperial regime of signs of the State apparatus capitalism becomes the line of flight that is charged with everything that cannot recharge the despotic centre; that becomes a negative line of flight heading into the desert and therefore allowing the circular, infinite movement of State signifier chains to continue.

The decoding of the axiomatic of capitalist globalisation works via connecting 'alien' unfamiliar products to everyday objects and thereby rendering them 'local' and familiar; decoding any imaginary of cultural, religious, and other differences in the process. Items such as '[j]unk food, toilet items, clothes, fashion, music, condoms, credit cards' were connected with each other and turned into the familiar – paradoxically allowing for 'a journey beyond familiar boundaries' through that very process (Nagaraj, 2012: 303). At the same time, the State has enabled new striations that cut apart the immanent connections of the protesters-prison-spade-ballot-box-war-machine:

> The primacy of the ballot box has now led to the marginalization of the other two [the spade and the prison]. Certain forms of connectedness . . . have disappeared from the body politic. Political parties have become election machines; all other structures, realms, and activities are subservient to success at the polls.
>
> (ibid., 301)

In that move, everyday globalised, decoded objects became confined to an isolated economic sphere, after they had previously been part of the body politics; as have been demands for justice and ethical considerations (ibid.). Indeed, the infamous argument that is so often made about those in poverty being unable to afford ethical arguments around issues such as biotechnology precisely relies on this logic. It is on the basis of the decoding of the capitalist globalisation axiomatic that pro-biotech-converted activist Gail Omvedt (1998) can argue that farmers 'grow cash crops because they have to sell in order for their families to survive: medicine, education and clothes, let alone radio, television and computers, are not subsistence products of the field.' This quotation is a fascinating example of the success of the interplay of the axiomatic and the State apparatus: goods such as medicine, education and clothes – the necessity of which can make them be the basis for emancipatory, anti-capitalist, political claims – are seamlessly connected to consumer goods that farmers (at least according to Omvedt) 'want' because '[t]hey want to be part of the modern world as much as everyone else wants to be' (ibid.).

This is why the protests of the KRRS against KFC had failed. There was no significant response to their protests and the consequent imprisonment of their leaders, and in the elections that followed the leading activists lost all of their seats except one (Nagaraj, 2012: 300–1). According to Nagaraj (ibid., 303), everyday objects got lost because they had not been preserved 'in the salt of cultural and religious memories'. The destruction of cultural memory, of different 'ways of life' and 'relations to land' relate to what I have previously defined as the reduction of ontological incompatibility in the context of India's particular colonial and postcolonial history of state, economic and social development. It is this history that might have led to a particular openness to the decoding processes of the capitalist globalisation axiomatic.

However, for Nagaraj not all is lost: he (ibid., 304) points out that the hunger-based identity's 'stubborn . . . insistence on fighting for the survival of different and, maybe even highly anachronistic, forms of everyday life' is precisely what prevents globalisation from accommodating this form of identity. This stands in contrast to other identities, such as the Hindu nationalist one that has fully taken on board the consumer logic of globalisation and merely aims to substitute it with the culturally/nationally 'authentic' (ibid., 292). Because globalisation relies on making the unfamiliar familiar by creating connections to the everyday, drawing on an entirely different 'authentic' traditional culture and way of life that, though imagined, continues to be a weapon in relation to an axiomatic that works by integrating resistance rather than opposing it head-on. 'Hunger creates a . . . kind of phantom which the rationalism of a market-machinery-driven universe cannot comprehend'; and in that 'lies the destructive powers of hunger-based identity' (ibid.). In other words, protests such as the KFC one aim for a reterritorialisation that invokes an essentialised binary between the 'modern' and the 'global' on the one hand, and the (imagined) 'traditional' and 'authentic' on the other. But it is precisely that move that contains the seeds for a possible destruction of the opposed logic as such.

It is at this point that I want to draw on the last of Deleuze and Guattari's four regimes of signs, which, as I will show, precisely encapsulates the potential of the hunger-based identity that Nagaraj identifies. The fourth regime of signs is what they call the 'postsignifying' one, which comes into existence by turning the negative line of flight that the State apparatus charges with everything that resists the latter's logic into a positive one. The line of flight of the despotic signifying regime now becomes 'a positive sign, as though it were effectively occupied and followed by a people who find in it their reason for being or destiny' (Deleuze and Guattari, 2004: 134). It becomes a 'positive line of [their] subjectivity, [their] Passion, [their] proceeding or grievance' (ibid., 135). While the war machine abolishes the imperial despotic line of flight and turns it back against the State, the postsignifying, subjectifying, passional line of flight turns it into a positive 'being' *via* subjectification. However, as Deleuze and Guattari emphasise, through subjectification the line of flight is yet again transformed into strata that at the end of the day 'repudiat[e] the positivity' and divert absolute deterritorialisation (ibid., 148). Though the postsignifying regime of signs is *different*, it is still a stratum that once again needs to be opened up to destratification and the play of becoming.

Conclusion: decolonising anti-GMO activism

Who's speaking? As this chapter has shown, the Indian Bt cotton controversy features some profound silences that counter all the noise on the surface of the debate. There is the silence of the nonhumans – Bt cotton, *Desi* cotton, the wind, the soil, the bollworm – that are mostly only spoken for from within fixed, given frameworks of understanding, with no space for an agency that goes beyond or that can challenge these frames. There is also the silence of the ones that Spivak might call subaltern – the landless migrant labourers, the small peasants – that is both a literal silence (at least in relation to the landless labourers) as well as a more general epistemic silence. This is a silence that has its roots in British colonial rule, but that has carried through into the building of the modern postcolonial Indian state, and that in the last decades has found new expression in economic liberalisation and an opening to the forces of globalisation. It is a silence that has resulted from the eradication of ontological alternatives to what Latour calls the two house-logic of the natural and the social, according to which science is the to-be-privileged body of knowledge that allows us insight into nature, that provides us with enlightenment and that is the basis for societal progress. As I have pointed out in this chapter, it is the embrace of this ontology that links the silence of the nonhuman to the silence of the subaltern, and that requires in response both historical analysis and a reflection on differentiated subject positions (as Spivak maintains). Moreover, it also requires a break with the ontological distinction between human and nonhuman. I have argued that in order to get at an ontology that is *different*, we need a new metaphysical understanding of the world. Deleuze's metaphysics of transformation points us to a (virtual) state of being that lies beyond that which is directly sayable or visible, and that connects to already-existing 'outside'-of-the-modern cosmologies that give a different ground to the relations of humans and nonhumans. Deleuze's metaphysics is helpful for making this connection because it understands concepts as created, unrepresentable 'beings' that in themselves point to a reality that goes beyond hierarchical stratifications characterised by identity and representation. Understanding concepts in this way allows the accessing of reality in a non-representable way, which avoids the danger of epistemic violence that *representative* concepts inevitably harbour. In the last section of the chapter, I have used several of Deleuze and Guattari's concepts – the various regimes of signs, the State apparatus, the axiomatic and the war machine – in order to reassemble the Bt cotton controversy in such a way that a move beyond the two-house logic became conceivable. It has led me to argue that a connection can be made between State apparatus-based agricultural striations and Man/gene (cf. Chapter 1), and that the capitalist axiomatic of globalisation has destroyed the traditional Indian democratic body politic by connecting unfamiliar nonhumans to the everyday. I have used the concept of the war machine in order to extract the prison-spade-ballot-box-assemblage that was able to connect humans and nonhumans beyond and outside of State/Man/gene and that, as a countersignifying regime, had been able to disrupt the despotic, imperial, signifying regime of the State apparatus. I have also shown that the interplay of the axiomatic and the State apparatus managed to destroy, through segmentation, this war machine's smooth space. But I have also pointed out, based on Nagaraj (2012), that the 'hunger-based identity' that we can find in the protests against globalisation, in which subjectification processes that are specific to the postsignifying regime of signs take place, contain a challenge to this interplay – precisely by drawing on an imagined essential traditional identity of rural India that cannot be accommodated by the forces of globalisation – though in the end, subjectification itself is yet again another stratum that needs to be moved towards destratification.

Nagaraj's reflections on farmers, globalisation and identity in India were written in 1996, two years before the start of the protests against Bt cotton field trials in which the KRRS was one of the leading players. It could be argued that the arrival of Bt cotton has since managed to release some of the destructiveness that Nagaraj has ascribed to the hunger-based identity; via enabling the forming of new assemblages that turned what forces of globalisation had made familiar back into the realm of the unfamiliar. As Visvanathan and Parmar (2003) maintain, some of the early traction of the KRRS anti-GMO protests goes back to the 'technological misconception' of the so-called terminator gene that I have already referred to in the beginning of this chapter (cf. Egovora, Raina and Mantuong, 2015: 107). Citing the example of a radio show made about the protests, Visvanathan and Parmar (2003: 2719) explain how the terminator gene (that from a merely 'factual' perspective did indeed not exist)[8] became connected to farmers' suicides as well as the potential need to sell off one's kidneys. This happened when the story of a poor farmer who had been tricked by a salesman over the quality of cotton seeds and felt consequently forced to sell his kidney to make ends meet was brought together with anti-GMO protests without any 'factual' connection between the two (ibid.). It is obviously impossible to draw a line between the early anti-GMO protests and the more recent actions of the Indian government that resulted, among other things, in the indefinite moratorium that was placed on GM crops in 2013. But as more recent studies have pointed out, there is nowadays a strong awareness of the links between neoliberal economic policies, globalisation, colonialism and internal problems of India, although the focus on science continues to be the dominant line of reasoning (see e.g. Egorova, Raina and Mantuong, 2015: 119).

The interesting question is whether the arrival of the GMO in India enabled the forming of an assemblage against the capitalist globalisation axiomatic that could ultimately be more successful than the one that was formed around the chicken in earlier years. The essentialisation of the GMO as the 'alien' might precisely be able to break what Nagaraj (2012: 303) calls the 'grow[ing] from strength to strength' of a globalisation that works through 'obliterating the differences between the [unfamiliar and familiar].' In this context, the question whether it is really 'true' that GMOs are inherently harmful, or whether there is really a 'factual' connection between suicides, kidney sales and the terminator gene is irrelevant. Moreover, the connections that are made between humans and nonhumans in these assemblages are not simply the result of (intentional or unintentional) strategies or tactics, but do *become*, and in this becoming they develop the power to break up existing striations of power and domination. Only if this is recognised are we able to not just develop new ways to evaluate the failure and success of anti-GMO activism in India and to build new strategies, but are also able to create new statements that move beyond an ontology that can only make sense of nonhumans as passive objects; an ontology that is deeply engrained in the very relations of power and domination that we want to contest.

Nevertheless, the problem of anti-GMO activists such as Shiva is that they continue to make too rigid a distinction between the 'alien' GMO and the 'familiar' traditional way of farming and living. This distinction is made from within a pre-given framework that leaves little agency for both nonhumans and (subaltern) humans to resist the questions they are asked – even if in the reality that lies beyond these frameworks, the agency *does* become manifest, as outlined above. With their ongoing fixation on the ontological properties of the GMO, their holding on to the idea of scientific 'facts' and the affirmation of a traditional, familiar identity of Indian agriculture, Shiva and other anti-GMO activists risk a constant cutting down of the nonhumans and humans that have found a voice in new, disruptive assemblages to a form that again correlates with the apparatus of State/Man/gene. Indeed, the ongoing emphasis on familiarity (in relation to traditional ways of life, to nature, to culture) once again opens a window for the forces of the axiomatic to accommodate resistance. Even if we go along with Nagaraj in maintaining that there is a destructive force in insisting on an imagined traditional way of doing things radically differently, there is still a case to be made for leaving space for unfamiliarity, dissonance and discomfort; for example by advancing a different understandings of nature/nonhumans as unfamiliar and (to put it in Deleuze's words), 'monstrous' in their difference (cf. Chapter 2). Solely invoking 'feel good' sensations about 'natural' nature and 'natural' nonhumans is inextricably related to the way that Indian anti-GMO activists such as Shiva make it too easy for Northern environmental movements and activists to 'feel good' about their own way of doing things (cf. Chapter 4; Braun, 2002).

As pointed out in an open letter from the Wretched of the Earth bloc, which fights for a decolonisation of the climate movement, the 2016 London People's Climate March for Justice and Jobs was a manifestation of ongoing colonial violence. Indigenous groups that had been invited to attend were put within a group of people dressed in animal headgear, instead of having been given a place at the front of the march as previously agreed. Banners of Indigenous communities that contained messages such as 'British Imperialism causes climate injustice' were covered in order to promote those that featured a more 'positive message'; one that was more palatable to the policy-makers to which the march wanted to appeal (The Wretched of the Earth bloc, 2016). Decolonising environmental movements can cause discomfort, alienation and disruption – but it is nevertheless crucial. Decolonising the anti-GMO movement might need to move beyond Shiva's invocation of global citizens who should connect with each other (see e.g. Shiva, 2006: 22), framed in a language and terminology that is recognisable to those squarely sitting within conventional modern ontology. Indeed, Shiva might have to take her invocation of the feminine principle herself more seriously – recognising the need to move towards a more profound and holistic challenge of the apparatus of Man/gene/State.

Notes

1   Though admittedly the analogy only goes so far. A 'weapon' is used to end and destroy what was before (or to threaten termination), while for Latour the aim of a proposition is to open up discussions instead of closing them down.

2 It also needs to be added that the research project that Carro-Ripalda and Astier's article emerged from, *GM Futuros*, was about smallholder farmers' voices and arguments in GM controversies in altogether three countries: Mexico, Brazil and, indeed, India (see Macnaghten, Egorova and Mantuong (eds), 2015). I will draw on the results of that project in relation to India later on in this chapter.

3 I have taken the liberty to merge what Deleuze and Guattari say about the creation of new statements towards the end of the plateau (2004: 162–4) with their elaboration on the relation of linguistics, abstract machines and machinic assemblages (ibid.: 156–60). However, their analysis in this plateau is far more fluid and less categorical and systematic than it appears at a first glance (this fluidity being in line with their general take on the function of concepts in philosophy). My depiction of their understanding of statements here is in many ways more *creational* than *representational* (which, for them, is the only way of doing philosophy anyway). It is, I would argue, a creation that helps me to move 'thought' for the purpose of this chapter.

4 Deleuze and Guattari (2004: 131–2) emphasise that the presignifying regime of signs is by no means to be understood in a historical evolutionist sense (as that which is somehow 'pre-modern'). They are at pains to point out that they do not want to do history, and that their insistence on all regimes of sign existing in mixed modes (even if one is dominant) is central for getting away from an understanding of time as chronological. Moreover, their categorisation of regimes of signs can once again only be understood in relation to their understanding of philosophy as *creating* concepts, which means that their 'categories' cannot be understood as rigid and exclusive.

5 Unfortunately, I myself have only become aware of this gap towards the very end of this book project.

6 With the term 'silence' I do not mean a literal silence (indeed, there is research that has engaged with voices of smallholder producers in India in relation to Bt cotton; see for example Egorova, Raina and Mantuong, 2015, and Glenn Stone's various publications, e.g. 2007), but a more profound epistemic silence.

7 For this it is also important to understand that prior to colonialism the Indian economy has not relied on agriculture, but that a majority of its population was forced into agriculture when its manufacturing industry was systematically destroyed by the British East India Company. See Tharoor, 2017: 7.

8 As Shah (2008: 437) points out, Monsanto's problem was precisely the ease with which the Bt cotton variety could be replanted, because the transfer of the Bt effect was guaranteed by simply crossing the existing Bt male with another hybrid female.

4 Travelling 'worlds'

The protest of the Intercontinental Caravan

According to a 2008 article in *The Guardian*, the anti-GM movements of the new millennium are not directly comparable to the ones we have seen at the beginning of the development of GE technology in the 1990s (at least in the Western world) (Sample, 2008). This, the author (ibid.) argues, is because they have found new strength in linking up with one another across national borders through embracing the concerns of the Global Justice Movement (GJM). As previously argued, concerns about corporate domination and global neoliberal governance have always been at the heart of anti-GMO resistance in the so-called Global South, while concerns about human and environmental health have dominated early anti-GMO resistance in Europe. Today, GE is primarily discussed in relation to issues of global justice and the need for economic and political transformation around the world. Regional or national fights against particular agricultural policies in countries such as India have managed to gain international recognition and receive expressions of solidarity precisely because they have linked those to a critique of global neoliberalism (Featherstone, 2003: 407; cf. Chapters 1 and 3).

While Chapter 2 has focused on late 1990s/early 2000s European anti-GMO argument (particularly in relation to science) and Chapter 3 on the Bt controversy in India in relation to the voice of smallholder producers, this chapter will engage more explicitly with the connections that have been formed between different movements across the world under the umbrella of the GJM. It will address the practices of protest and connection with a particular emphasis on the problems that come to the fore from a decolonial perspective. As I will argue, focusing primarily on what sort of global structures and institutions need to be critiqued and attacked leads activists to strive for sameness in intention and strategy, as well as harmony and positivity among those participating in protest movements. This ignores the existence of internal structural hierarchies and too easily leads to the reproduction of colonial forms of oppression and injustice. Though this argument has already been repeatedly made in relation to the GJM (particularly concerning the annual World Social Forum (WSF) meetings), my aim in this chapter is to work out the deeper philosophical implications of these problems, which are (once again) related to the dichotomies prevalent in the modern/colonial logic as such.

The case that will be used for this kind of analysis is a historical one. In May/June 1999 several hundred representatives of the Karnataka Rajya Raitha Sanga

(KRRS – Karnataka State Farmers Organisation) that I have already engaged with in Chapter 3 travelled, together with other Indian farmers and activists from all over the world, around Europe with the so-called 'Intercontinental Caravan' (ICC). The objective was to protest at the sites of some of the world's most powerful organisations and companies against the impact of neoliberal globalisation, particularly in relation to agriculture and the technology of GE. The event preceded the 'official' start of the GJM, which is often taken to have been the so-called 'Battle of Seattle' protests in November 1999, but in its composition, 'modus operandi' and objectives it already contained many prominent GJM features. It was set up as a travelling protest that some of those who had accompanied it described as carnivalesque and theatrical (see e.g. Madsen, 2001). Additionally, through the sharing of food, shelter and company for several weeks in concrete spaces, it also already employed (though not necessarily intentionally) some of the so-called 'prefigurative strategies of transformation' (i.e. *living* the transformation) that have become significant in more recent anti-neoliberal protests. While the 'network' was a prominent concept to describe earlier GJM manifestations, many of the Occupy movements returned to emphasising the need for concrete localities of/in protest.

My aim in this chapter is to step out of the binary network/fixed space logic altogether and start from within a different frame. From a decolonial perspective I will engage with both enthusiastic affirmations of the ICC and with the arguments of its critics, and I will argue that both enthusiasm and dismissal ignore the nature of colonial violation. In other words the *logic* of oppression, which lies in the very categories that many (particularly European) activists have at hand to mark out the problem they want to address, to design protest strategies and to judge the success or failure of a protest, is neglected. After analysing the problems of the ICC from that point of view, I will then go on to ask about its productivity from a position that lies outside of these given categories, drawing for that purpose on María Lugones's book *Pilgrimages/Peregrinajes: Theorizing Coalition Against Multiple Oppressions* (2003). Based on this, I will argue that the liberating potential of the ICC lies in the way it provided its participants with an opportunity to become what Lugones (ibid., 7) calls 'faithful witness[es] . . . against the grain of power': pilgrims and streetwalkers who connect to each other in a spatiality of relational concreteness that is neither the one of the network nor the one of fixed space. Rather it is something 'impure' in-between, kind of both, or neither-nor, that, as I will argue in the last section, is decisively enabled through nonhuman participation.

A politics of network? The Global Justice Movement

In the 1960s and 1970s new social movements emerged that, instead of focusing on traditional concerns over economic, labour-related issues, turned their attention towards themes such as civil rights, sexual identity and environmental concerns. Since then the academy has been theorising social movements within a 'recurring dichotomy': it has understood them either in 'instrumental terms' (as rationally motivated and organised) or in 'expressive terms' with an emphasis on identity, the symbolic or communities (McDonald, 2006: 17). Theorists have consequently determined the strength and success of a movement by either their commonality of interest, their ability to formulate a coherent strategy and the existence of sufficient resources (in the instrumental approach), or the existence of shared identity traits, such as sex or ethnic background (ibid; O'Neill, 2004: 239).

This strong focus on the need for a shared identity, a set of shared interests and grievances, and/or a clear political objective and strategy has initially led to scepticism about the chances of success of what was variably called the 'Anti-Globalisation Movement', 'Alter-Globalisation Movement', or 'Global Justice Movement' (I will stick to the latter term in this chapter). The emergence of the GJM is often dated back to the famous 'Battle of Seattle' in 1999, in which thousands of people from around the world protested against and finally managed to shut down the latest WTO round of trade negotiations.[1] Given the dispersed nature of the many movements that have participated in the GJM, with different interests, backgrounds, grievances and objectives, some researchers have initially doubted the impact that the GJM could have. Neil Smith (2000: 4), for example, maintains that the strength of the 'Battle of Seattle' – 'internationalism, resoluteness and breadth' – also made up

> its central weakness insofar as the convergence of such an eclectic political grouping is not dependable without a sharpened political focus and enhanced organizational power.

A decade later, Geoffrey Pleyers (2010: 11) continues to argue that the capacity of the GJM to 'act' depends on it having 'a coherent logic of action'. A 2012 study that draws on data and texts from 45 organisations that are linked to the WSF (which is an annual meeting of organisations and movements connected to the GJM that first met in Brazil in 2001) concludes that it precisely has that capacity. Manfred B. Steger and Erin K. Wilson (2012: 440) maintain that what they call 'justice globalism' is today indeed 'a congealing political ideology offering an alternative conceptual framework for collective political action.' Moreover, they (ibid.) maintain that this 'ideology' is not abstract, but contains 'practical policy alternatives to those currently touted by market globalists' (cf. Steger, Goodman and Wilson, 2013).

But whether the GJM is successful according to the parameters of conventional social movement theory or not, many scholars have since argued that we need to focus on the way the GJM fundamentally challenges old categories and concepts. As Graeme Chesters and Ian Welsh (2005: 187) argue, the GJM embraces the fact that there are *both* commonalities *and* differences within and between them. The GJM 'subject' does not seem unified or even willing enough to negotiate particular interests, identities and objectives in relation to (global) regimes of governance in the way that conventional social movement theories would expect it to do (cf. Wilkin, 2003: 87; Chesters and Welsh, 2006). Chesters and Welsh turn to the work of Deleuze and Guattari to find new concepts that enable them to grasp the decentralised and networked character of the GJM. At first glance, their drawing on Deleuze and Guattari is similar to the way I engage those authors in Chapter 3 of this book: Chesters and Welsh (2005) use the concept of the war machine in order to argue that the GJM has become 'an anti-capitalist attractor within global civil society' that allows for the forming of new assemblages through 'spaces of encounter and deliberation' (ibid., 188). The 'antagonistic quality' of GJM resistance practices does not just lie, they argue, in their 'acts of political and economic contestation', which could indeed be explained from within conventional frameworks, but also in 'its capacity for cultural intervention and experimentation' (ibid.). In other words, resistance is not just found in the claims that are made and the policy proposals that are put on the table, but in the deployment of a different subjectivity that fights (neo-)liberal regimes by contesting the (liberal) model of agency that underlies them: 'poetic utterances' are preferred to 'political rhetorical', and 'carnival rather than collectivism' becomes 'its modus operandi' (ibid.). What makes the multiple movements develop 'shared understandings' is the encounter and interaction that takes place in 'spaces of intensive networking' (ibid., 192).

For Michael Hardt and Antonio Negri (2000, 2006), this different 'modus operandi' of contemporary social and political struggles has the capacity to challenge contemporary global power relations at their core. In our globalised world, they argue, capitalism is increasingly dependent on 'immaterial labor' that produces 'new subjectivities' and 'forms of social life', resulting in new, immanent, biopolitical forms of domination ('Empire') (Hardt and Negri, 2006: 66; cf. 2000). However, the 'biopolitical process' does not simply reproduce capital as social relation, but also provides 'the potential for an autonomous process that could destroy capital and create something entirely new' (Hardt and Negri, 2009: 136). 'Immaterial labor', embedded in immanent 'networks based on communication, collaboration, and affective relationships' leads to the emergence of a new force of resistance: the 'social composition of the multitude', which is different from both the concept of 'the people' (as unitary entity) and 'the masses' (characterised by indifference) (Hardt and Negri, 2006: 66, xiv). For Hardt and Negri, the multitude is characterised by openness, inclusivity, and the capacity to 'communicate and act in common while remaining internally different' (ibid., xiv). This way it manages to encompass a variety of movements and events, from trade union strikes to the GJM (ibid.).

There are various fundamental problems with this sort of argument. I have already pointed out in Chapter 3 how the application of Deleuze (and Guattari's) concepts to actual phenomena disregards Deleuze's philosophical objective of moving away from representation. The lack of proper engagement with Deleuze's metaphysics leads to the instalment of yet another set of binaries ('Empire' versus the 'multitude') and a hyperbolic enthusiasm for the side of the 'resisters'. The result is a fundamental failure to come to terms with, let alone overcome the dogmatic image of thought. Despite Hardt and Negri's emphasis on the 'internal difference' of the multitude, their understanding of difference adheres to the logic of what Deleuze (2004: 64), drawing on Hegel, calls the 'beautiful soul':

differences, for them, are 'respectable, reconcilable or federative'; leading in the end to a forming of a Whole (*the* multitude) the emancipatory potential of which is taken for granted (cf. Coleman, 2015a; Coleman and Rosenow, 2017b). Concepts such as 'multitude', the 'war-machine' or the 'rhizome' are inserted into what is still a modern/colonial epistemic and conceptual framework: Chesters and Welsh, for example, continue to use the concept of 'global civil society' (in which the GJS acts as an 'anti-capitalist attractor') as if it were an unproblematic one, and Hardt and Negri seem to take as given (prior to analysis) what movements are sufficiently 'emancipatory' in order to be included in the resisting force of the multitude. Theorisation takes place from the position of 'the ideal observer', who 'must himself be pure, unified, and simple so as to occupy the vantage point and perceive unity amid multiplicity' (Lugones, 2003: 129). The (colonial) vantage point easily leads to abstracting resistance to a 'thing' that is deduced from pre-given logics of domination, instead of regarding it as 'cultivated, through difficulty and elaboration, from a subaltern positionality' (ibid., 23; Lugones here refers to Ranajit Guha's work on peasant uprisings). In other words, the focus on *one* (unified) logic of resistance and domination is in danger of making invisible the actual lives, socialities and struggles of 'thick' oppressed individuals on the ground.[2]

Since the early 2000s various critics have started to unpick the workings of hegemony and domination inside the GJM itself (see e.g. Worth and Buckley, 2009; Escobar, 2004; Sullivan, 2005; Rosenow, 2017). Paul Routledge, Andy Cumbers and Corinne Nativel (2006) have coined the term 'imagineers' to describe how certain dominant movements and individuals shape and dominate the 'imaginary' of a GJM network (cf. Routledge, 2008). Escobar (2004) has pointed out how the WSF continues to be dominated by hierarchical organisations such as trade unions and political parties. From a decolonial perspective, it is striking that neither Chesters and Welsh nor Hardt and Negri seriously engage with the problem of coloniality in their depiction of contemporary global power relations. As Mignolo (2007) emphasises, alternatives need to be found not just to neoliberalism – or even capitalism – but to Western rationale and practice '*in toto*, that is, liberal and neo-liberal but also Christian and neo-Christian, as well as Marxist and neo-Marxist' (ibid., 456; emphasis in original). Instead, both Chesters/Welsh and Hardt/Negri continue to draw on a European body of thought in order to mark out and contest contemporary forms of capitalism.

Potentially also in response to some of the critiques of the earlier hyperbole about the emancipatory potential of global 'networked' resistance, more recent protest movements and events – the 2011 Arab Uprisings, the Spanish Indígnados protests, and the Occupy movements, to name only a few – placed renewed emphasis on the importance of 'local' space. Prefigurative strategies for social change aimed to put into practice desired social transformations in a confined space and time (see e.g. Leach, 2013; Hardt and Negri, 2011; Prentoulis and Thomassen, 2013). Sam Halvorsen (2012), who participated in the Occupy London movement in 2012, explicitly challenges the 'logic of the network' that had come to dominate interpretations of the pre-Occupy 'global movement for socio-economic justice' (ibid., 431). He points at the 'reluctance' of the Occupy movements 'to create the infrastructure for a global networked movement', and traces this reluctance back to 'a renewed commitment to non-hierarchy that puts into question both the scale of organising and the function of global networks' (ibid.). Instead of embracing the spatiality of the network, we now witness, according to Halvorsen (ibid., 432), a return to 'the agency of territorially based social movements, against the trend to always extend outwards and build connections'. Protesters build a counter-temporality that uses 'fixity and territory as a weapon' instead of emerging from and going with the 'fast-paced rhythms and flows' of complex contemporary relations of power and domination (ibid.). However, this interpretation of more recent social movements as somehow *returning* to fixed and local space once again bows to binary 'either-or' thinking. It is *either* fixed space and time, *or* networks, flows and complex encounters.

'World'-travelling and multiple selves: an introduction to María Lugones

María Lugones's work is often discussed together with the work of Gloria Anzaldúa (1987), Cherríe Moraga (1983), Chéla Sandoval (1995, 1998) and other so-called 'Latina feminists' (Ortega, 2001) or, as they are also sometimes called, 'U.S. Third-World Feminists'.[3] Based in the US but having their roots in Latin America, Latina feminists develop feminist theorisation out of the concrete life experiences of those who (like themselves) are 'condemned to live the life of the "world"-traveler' (Ortega, 2001: 3), and who are therefore in a position to 'convey [. . .] meaning against the oppressive grain' (Lugones, 2003: 7). Lugones defines the 'world'-traveller as the one whose being is placed (in contrast to the existence of the one who occupies the dominant position) across the borders of multiple, often contradictory 'worlds', and who thereby qua existence contests the common sense of coherent, transparent individual identity.[4] A 'world'-traveller exists in multiple 'worlds' as multiple selves, featuring 'a multiple sensing, a multiple perceiving, a multiple sociality' (ibid., 11):

> When I think of my own people, the only people I can think of as my own are transitionals, liminals, border-dwellers, world-travelers, beings in the middle of either/or. . . . So as soon as I entertain the thought, I realize that separation into clean, tidy things and beings is not possible for me because it would be the death of myself as multiplicitous and a death of community with my own.
> (ibid., 134)

Importantly, for Lugones all the 'worlds' that the multiple selves occupy are 'actual' and lived – none of them are possible (or 'virtual', as Deleuze would have it), and they have a particular concrete spatiality. A 'world' in Lugones's sense is 'inhabited at present by some flesh and blood people', though it can also encompass people who 'the inhabitants of this "world" met in some other "world" and now have in this "world" in imagination' (these people might even be dead in 'this world') (ibid., 87). Lugones's emphasis on the lived and on actual experience shows that she is close to certain traditions of phenomenology, insofar that she 'attempt[s] to close the gap between theory and practice, between how we think of the world and how we live it' (Ortega, 2001: 3). She could also be seen as related to feminist standpoint theorising, insofar that her philosophy privileges the experiences of oppressed existence. However, she herself argues that she has been most influenced by black feminists, who have for a long time critiqued the 'boomerang perception' of (hegemonic) white feminists who are only able to conceptualise 'black difference' by comparing it to themselves as standard.[5] White feminists, Lugones (2003: 68) argues, have come quite some way in recognising the problem of difference, but most of their theorisation remains untouched by that problem: it continues to be based on the experiences of white women which are taken as the norm (ibid.). Lugones's concepts of 'world'-travelling, multiple selves, curdle-separation and others (which I will explain at a later point in the chapter) aim to counter this tendency, by writing general philosophy from a starting-point that is grounded in the lived experience of 'intermeshed oppressions' (ibid., x). Though today Lugones might be more known for her intervention into decolonial thought (see e.g. Lugones, 2007, 2010), in this chapter I will mainly draw on her earlier work that has been brought together in her book *Pilgrimages/Peregrinajes* (2003). Here, Lugones emphasises the resisting potential of what she calls 'active subjectivity', which is a subjectivity that neither draws on the subject as individual nor on the subject as 'intentional collectivity of collectivities of the same' (ibid., 6): 'active subjectivity . . . is adumbrated to consciousness by moving with people, by the difficulties as well as the concrete possibilities of such movings' (ibid.). Similar to some of the theorists that I have referred to in relation to the GJM, Lugones emphasises the importance of not falling back 'into a politics of the same, a politics that values or assumes sameness or homogeneity' (ibid., 87). But her concepts of multiple selves and 'world'-travelling, including the concrete spatial dimensions that this involves, allow her to avoid the sort of enthusiastic hyperbole of the 'politics of connectivity' literature that I have critiqued in the previous section. Lugones realises that a 'person may be both oppressed and resistant and act in accordance with both logics' (ibid., 13). Moreover, the multiplicity of the self and the 'worlds' that selves travel lead to '[e]xpressive gestures, acts, movements and behaviours' that might make sense in *one* world, but are 'incommunicative' in others (ibid., 25). That said, none of this precludes for her the possibility of moving and resisting together: in contrast to much poststructuralist literature, she actively embraces, *from* the position of the oppressed 'world'-traveller, the 'liberatory possibility' (ibid., 58).

Lugones's take on multiplicity, heterogeneity and spatiality goes against the (colonial, hegemonic) desire for sameness and homogeneity. It grounds central concepts in the to-be-privileged experience of the oppressed and allows for a critique of the coloniality of much existing 'global' activism (particularly in the environmental-agricultural realm). In the next section, I will give depth to this critique by engaging with the historical case of the 'global' protest of the ICC. However, Lugones's affirmation of the need for wholesale 'liberation' also allows me, as I will show, to formulate concrete suggestions of how to move forward towards a different transformative political practice.

'In Asia great leaders are expected and revered': the colonial logic of the Intercontinental Caravan

In this section I will critically engage with the colonial logic that has, I argue, (implicitly) underlain the organisation and practice of the ICC, particularly in the way of which the encounter between participants from different historical, cultural and geopolitical backgrounds has been coordinated and made sense. The colonial logic that I will identify becomes particularly manifest in the desire for homogeneity, the striving for 'purity' in aims and forms of protest, and expectations of an essentialised 'peasant' identity. This, I will show, can be traced back to a lack of reflecting how coloniality has shaped, and continues to shape, the relationship between colonisers and colonised, and their perception of each other.

From the point of view of many of the Indian ICC participants, the constitution of a global 'we' as a political subject presupposes a particular essential Indian 'we', featuring a particular (national) identity and culture. Linked to the idea of what it is like to be a farmer in rural India – which Nagaraj (1996) calls, as I have already pointed out in Chapter 3, the 'hunger-based identity' – it leads to an emphasis on the possibility of a life that can be radically different from the one brought by globalisation and 'the West'. A strong distinction is made between an 'us' (the Indian farmers) and a 'them' (globalisation, MNCs, the Indian state, Western states, international organisations). The strong ideal of autonomy leads to an approach to the global that is supposed to work bottom-up, meaning that independence at the local level, manifested in autonomous decentralised communities, is a necessary precondition for cooperation and solidarity at other (national, global) levels (Nanjundaswamy, 1999; 2003).

The idea of autonomy and decentralisation goes hand in hand with the ideal of authenticity, which Indian ICC participants do not only claim for themselves, but which is also attributed to them as cultural 'other' and 'true' voice of grievance from their European allies. This led to internal controversies within the ICC because Indian participants were accused of not being 'authentic' smallholder peasants. Despite agreeing on the grievances that needed to be addressed, such as the power of MNCs, the opposition to GMOs and the critique of large-scale unsustainable farming, there were quarrels among Indian activists about how to organise the protest – particularly between the KRRS and Vandana Shiva. In an email Shiva claims that she was approached by a delegation of farmers from Karnataka, some of them members or former members of the KRRS, to warn her about the ICC:

> They wanted me to know that since most Indian farmers are buried under debt and thousands have committed suicide over the past year due to indebtedness, no farmer can afford to pay Rs 35,000 for travel to Europe. Most of the so

called 500 farmers who will be travelling to Europe as part of the 'caravan' are basically bank officials, pesticide and seed agents, and commission agents.
(quoted in Madsen, 2001: 3739)

Similarly, an anonymous European critic points out that organisers in Europe suffered from the expectation that 'visitors would be true peasants', while in reality most participants were 'middle-class small farmers, some of whom were here looking for work, business opportunities or just a cheap package tour of Europe' (Anonymous, 1999: 28).

Concerns about the need for a coherent, essential and collective identity as well as coherent, unified intentionality (i.e. if one goes on a protest one cannot at the same time do sightseeing or look for business opportunities) are grounded in an understanding of oppression and self as unified and homogenous. By contrast, as already pointed out, for Lugones 'active subjectivity' that targets oppression does not presuppose coherent individual selves or collective intentionality. Resistance needs to be intentional, but intentionality is emergent and can only evolve when attention is paid 'to people and to the enormously variegated ways of connection among people' (Lugones, 2003: 6). This intense paying of attention to people in their complexity is only possible if we start to *sense* oppression, resistance and intentionality, rather than *make sense of it* monologically in and through the ways that we have been taught by the oppressor (and which we call *common sense*). Sensing resistance implies sensing people's 'desires, beliefs and signs' (ibid.) by starting to understand and actively shape oneself as multiple and as travelling 'worlds' (indeed, the more coherent we feel, the closer we adhere to the oppressor's logic). It implies that we stop valuing 'sameness' or 'homogeneity' (ibid.). While postmodern theorists, according to Lugones, fundamentally contest the 'politics of identity' and 'the political significance of groups' as such, Lugones's phenomenological approach leads her to affirm both identity politics and the need for groups as central to our experience of our 'world(s)' – but we have to understand both selves and 'worlds' as multiple (ibid., 142). This then allows us 'to stand in the cracks and intersections of multiple histories of domination and resistances to dominations' (ibid., 6)

Indeed, what is lacking in the way the ICC is both dismissed and celebrated in the sources I have at hand[6] is a closer understanding of the multiple historical logics of oppression and domination that I have worked out in Chapter 3, such as for example the factors that have led to the (epistemic as well as literal) 'silence' of Indian smallholders (or 'peasants'). Both European and non-European ICC activists and supporters (or critics) who have spoken out and/or written about the ICC seem to understand oppression as clear-cut and unified, so that one is *either* on the side of domination *or* on the side of the resisters. There was also the related expectation that understandings of nature and the right way to farm are the same among all protesters. As David Featherstone (2003: 415) who participated in the ICC points out:

[T]he ICC project . . . located the opposition to genetically modified organisms at the intersection of different routes of activity and experience. . . . The

ICC was not mobilized around a consensual version of what constituted 'the environment'.

European ICC anti-GM activism was grounded in ideas of purity that led to the collapsing of Indian farmer identity into nature, in typically orientalist fashion, implying that 'good' farming practices were associated with Indigenous 'primitive' agricultural tradition. By contrast, as already pointed out at various points in this book, Indian participants were more focused on opposing economic exploitation. Indeed, according to Featherstone (ibid., 415–16), many European activists appeared to be shocked by the 'productivist' tone of some of the Indian farmers, who were for example in favour of the use of pesticides on their fields. Once again, the intermeshed oppressions towards which the multiple selves of Indian ICC participants were directed were not recognised by their European allies: there seemed to be no understanding of how the enthusiastic belief in the possibilities of technological progress, science and agricultural productivity (resulting at least implicitly in an acknowledgement of nature as exploitable resource) has emerged from the foundations of an Indian postcolonial state that was established *in continuity* with the European paradigms that had been introduced during colonial times (cf. Chapter 3).

Lugones critiques the notion of purity in her chapter on 'mestizaje' (a term that usually refers to those Latin Americans who have a mixed European-Indigeno ancestry). 'Mestizaje' becomes Lugones's central metaphor for the need to understand resistance as 'impure', as something that is 'neither/nor', 'kind of both, not quite either', 'the middle of either/or', or 'ambiguous' (Lugones, 2003: 122). For Lugones, 'pure' separation (which means that what is to be separated has to 'split' in 'clean' parts) is part of the logic of domination, while the logic of resistance is one of impure 'curdling'. The 'impure', the in-between, is precisely what is 'unclassifiable', 'unmanageable', and therefore outside of the reach of a power that needs clear-cut parts in order to exercise control (ibid.). As I have shown at length in Chapter 2, this is the same logic that underlies the practice of GM. In line with the Central Dogma of mainstream molecular biology, the central assumption of GE is that an organism can be 'split-separated' (to use Lugones's terminology) into different parts of a controllable, unified whole. Unity can only be grasped from a distant vantage point that is based on the purity, simplicity and unity of the observing subject itself. The 'urge of control' that this position implies is 'conceptually related' to 'the passion for purity' (ibid., 127).

In relation to GE, this is precisely the position of the scientist, who is sufficiently distanced from his object and sufficiently 'pure' in his mind to suppress any complexity that goes along with 'sensuality, affectivity, and embodiment' (ibid., 130). Impurity is a central feature of the oppressed subject: a 'subject who is in multiplicity' and therefore lacks the 'unidimensionality' that is required for the vantage point of the dominant (ibid., 129). Though anti-GMO activists implicitly critique this logic of 'purity' and 'split-separation' when it comes to GE, they often embrace it when it comes to traditional agriculture, true and authentic peasanthood, and their general notion of nature. The logic of hybridity is rejected in favour of one that Lugones, in relation to how the Mexican/American is regarded by the white/Anglo-American, calls 'dual': the 'authentic Mexican cultural self' is a 'simple but stoic figure who will defend the land no matter what', and is contrasted with 'the American self' that is the opposite (ibid., 134–5). What is not sufficiently recognised is that the (colonial) logic of purity and split-separation underlies *both* the concept of GM *and* the concept of traditional versus modern identity.

The Indian farmer who is (implicitly) dismissed by the European anonymous critic as impure – insofar that s/he has not just come to Europe to protest – precisely occupies the position of the subject in multiplicity: s/he might want to be tourist, businessman/woman and protester at the same time, s/he opposes Monsanto and GMOs but at the same time embraces the use of pesticides in agricultural practice. By the same token, some of the Indian participants also failed to regard themselves as what Lugones calls 'thick' individuals, which are individuals who are aware that there are always aspects to them that make them different from other members of their group; leading to the existence of different 'needs, interests, ways' that are 'relegated to the margins in the politics of intragroup contestation' (ibid., 140). When an argument broke out about the 'imbalance' created by the very large Indian group compared to the other non-European ICC participants who felt marginalised, as well as about accusations of sexism directed at female participants e.g. expected to do all of the cooking, one of the Indian leaders urged everyone to stick together: 'After all, we are all Indians and we have come here as Indians' (quoted in Madsen, 2001: 3741). This made matters worse, precisely because NOT all of the participants were Indian, and it was replied that 'the ICC would fight any form of imperialism, including India's imperialism vis-à-vis Nepal' (cited in ibid.). Being Indian (and, by implication, being male) had become the central criterion for defining group membership, and 'thick' members of the group (such as Indian women or non-Indian non-European participants) were marginalised. Similarly, assumptions of transparent (= non-thick) sameness and homogeneity were also ascribed to Indian participants from the outside, having led to judgements that testify to the ongoing existence of the colonial logic in full blaze:

> *In Europe*, decentralisation and non-hierarchical organisation are an important part of our political consciousness, but *in Asia* great leaders are expected and revered.
>
> (Anonymous, 1999: 29; emphasis added)

The colonial oppressor looks out for a difference that is comparative to her/himself as standard, and that can ultimately be reconciled with the latter. In her book Lugones recalls an argument that she had with a white woman who happened to sit next to her on a plane, and who forcefully maintained that Latinos/as and black people in the US are not subjected to racism; that, instead, it is now recognised 'that we are all the same' (Lugones, 2003: 158). But then the conversation turned towards Lugones's fellow traveller's disgust of Black and Latino rap and the violence against women that it, she suggested, justified. Consequently, 'the stereotype', Lugones (ibid.) says, was brought 'right back to my face': '[the white woman] insisted that Latino and Black male rappers were, after all, different, brutal, animalistic, sexually violent in a way very much different from white manhood and from the violence of the system' (ibid., 159). What Lugones consequently calls the desire for '[s]ameness called for by narcissism' is also found in the judgement of the anonymous ICC critic: we are all the same, we stand in solidarity, but the moment you do not follow the boomerang of my perception, you are not only different, but monstrously so, as previously quoted statement manifests (ibid.).

At a first glance, one of the positive accounts of the ICC that has been written – European participant and organiser Katharine Ainger's contribution to the famous GJM publication *We Are Everywhere: The Irresistible Rise of Global Anti-capitalism* (2003) – seems to go against that logic, by celebrating the encountered difference of the non-European ICC participants as productive. However, Ainger's depiction of Indian farmers is in many ways as colonial as the previously mentioned comment of the anonymous European critic, as following quote exemplifies:

> I'm at the airport to meet a hundred Indian peasants from the KRRS off their chartered Russian plane. . . . They walk in the white sterility of the airport with a banner proclaiming their arrival, past adverts for corporate bank accounts, global financial services and consumer items for business travellers. Their contingent fills an entire baggage-reclaiming belt. Then, with Indian flags flying on their airport trolleys, they pour into Fortress Europe. . . . The following morning, I wake beside dozens of sleeping bodies, my hair filled with straw. . . . As I descend the wobbly ladder, the mist lifts off the surrounding fields, and I try to take in the improbable view below me. A rural Indian hamlet has been dropped wholesale into the German countryside. Hundreds of Indian farmers are wandering around in the chill of the early morning, many wrapped in the trademark green shawls . . . of the Karnataka farmers. A serene old man wearing kurta pajamas and plastic shoes in the centre of the courtyard takes a long, slow draw from a hookah pipe as large as himself, and still with its Aeroflot baggage-tag attached.
>
> (Ainger, 2003: 160–1)

The coloniality of Ainger's gaze manifests in her inability to have her own frame of reference challenged by the difference she encounters. Instead, Ainger takes everything she sees and hears as confirmation of her own take on the world and of what she wanted the ICC to be about. The Other – who is once again primarily be determined by her/his exotic otherness – cannot determine her/himself what the 'world(s)' is about, but simply fits right in. Ainger's understanding of difference is again the one of Deleuze's beautiful soul in which all difference is reconcilable and working in harmony (Deleuze, 2004: 64).

But that was not the only effect that the ICC had. Featherstone (2003: 415–16) also recollects conversations with European participants who found that their direct encounters with farmers from the Global South productively unsettled some of their strong views on nature and (anti-)development, and he regards these encounters as one of the major achievements of the ICC. Did the ICC provide the opportunity for a better understanding of 'the complexities of resistance to intermeshed oppressions' and for the creation of multiple sensing, perceiving and sociality after all (Lugones, 2003: 8, 11)? To find out, I will now turn to Lugone's concepts of pilgrimage and streetwalking, with a particular interest in the notion of space that is coming to the fore in both.

Pilgrimage and streetwalking: the decolonial option

Anti-structure and communitas

In this section I will show what kind of different (decolonial) perspective on the ICC (and by implication global protest movements more generally) emerges if we apply to it Lugones's concepts of pilgrimage and streetwalking. For Lugones, thinking about resistance to intermeshed oppressions as a pilgrimage allows us to move against structures of power that fragment and keep apart the existing 'manyness of worlds of sense' (ibid., 6). Viktor Turner (1974), whose work on pilgrimages Lugones uses for that purpose, understands the pilgrimage as a 'form of institutionalized or symbolic anti-structure' that 'breeds new types of secular liminality and communitas' (ibid., 182). Conceptually Turner draws for this argument on his research on transition rites in tribal societies, which, he argues, allow for the constitution of a state of liminality in which a group of 'novices' suspend for a short period of time the social structure that makes them 'social *persona* segmentalized into a series and a set of structural roles and statuses' and constitutes them, at least for the moment, 'as equals' (ibid., 201). Liminality therefore provides the 'optimal setting' for relations of 'communitas', which is the 'spontaneously generated relationship between leveled and equal total and individuated human beings' (in contrast to 'community', which is about 'a geographical area of common living') (ibid., 201–2). Turner maintains that liminality and communitas together form an 'anti-structure' in which 'all structural rules' are put into question and 'new possibilities' emerge (ibid., 202).

For Turner (ibid., 169–70), the communitas of the pilgrimage is a 'normative' one in which a 'social bond' is created between pilgrims and their hosts through a shared normative commitment. The social structure of kinship and neighbourhood is left behind in order to form relationships with 'the generic human "brother" and "neighbour" who might be anyone in the wide world but whom one should "love."' (ibid., 186) However, in contrast to transition rites, pilgrimages are still structured by religious systems that ground the normative commitment. The generic brother is still the one who shares a given system of beliefs, with the one who doesn't being excluded (ibid., 206). Turner argues that the simultaneously inclusive and exclusive nature of the pilgrimage manifests itself topographically: the religious activity is usually focused on sites that 'are themselves parts of bounded social fields' and of hierarchical, structured social systems (they take place at shrines, in churches, or temples), while the 'pilgrim centers' where pilgrims sleep, eat and rest are usually all-inclusive (ibid., 186). When a pilgrimage does not 'force' communitas ('forced' communitas characterised, for example, the Crusades) but 'operates within relatively wide structural limits', it offers an opportunity for groups and individuals to bind together in diversity and to 'overcom[e] cleavages' (ibid., 206).

Lugones employs Turner's concept of the pilgrimage in a non-literal, epistemic way. For her (2003: 61), liminality is not a temporal interval between structures, but a general 'social state' that allows for more fundamental structural critique. The 'limen' is constitutive of a self that finds itself targeted by intermeshed oppressions, and it demarcates the space between the different 'worlds' that the self occupies; allowing for the sociality of communitas. Anti-structure becomes an inherent feature of the multiple self (ibid., 61). On the pilgrimage 'selves, relations, and practices' can be thought 'as constituting space and time away from linear, univocal, and cohesive constructions of the social'. They can thereby 'loosen the hold of institutional, structural descriptions in the creation of liminal spaces' (ibid., 8). Desirable 'collective practice' involves a 'dialogue among multiplicitous persons who are faithful witnesses of themselves and also testify to, and uncover the multiplicity of their oppressors and the techniques of oppression afforded by ignoring that multiplicity.' (ibid., 62)

Given the set-up, objectives and modus operandi of the ICC, the question emerges whether we should not go back to Turner's concept of the pilgrimage in a more literal sense (being obviously aware that the ICC was not a completely literal – i.e. religious – pilgrimage); taking into consideration Lugones's understanding of the limen as a social state at the same time. As with other critical thinkers that I have engaged in this book so far, it is sometimes difficult to imagine what sort of concrete realities of resistance (over and above an individual engagement of the self with itself and its immediate others) Lugones wants to move towards. The ICC participants travelled around in a literal coach 'caravan' and temporarily occupied concrete roads, squares and other public spaces. They ate, slept and spent leisure time together and thereby created (or at least had the opportunity to create) space and time away from the structures they were part of at home. Local people were variously engaged as hosts, friends, foes, allies or spectators (Madsen, 2001: 3733). There was definitely a proclaimed shared system of beliefs among the participants that had made them commit to the journey and to other participants (whom they often had not chosen) as 'brothers', 'sisters' or 'neighbours' in the generic sense that Turner is talking about. There is both exclusion and inclusion, and often both work topographically. As Turner says, on the pilgrimage anti-structure is not as complete as it is in transition rites: structural divisions are not removed, but there is potential to take the 'sting' out of them, and to envision, in this moment, new possibilities. The decisive difference between Turner's description of the pilgrimage and the self-understanding of the ICC (as well as the self-understanding of social protest movements more generally) is that members of the former are usually very well aware and even accepting of existing divisions, while in the latter such divisions are often either denied or not thought about, and then decried when they come out. Moreover, those who participate in a political protest usually seek *solidarity* instead of *communitas*. While the former defines unity as that which belongs to an in-group that is opposed to an out-group (here are the protesters, over there the MNCs, international organisations and 'the West'), the latter's ambition is universality. Seeking solidarity against an enemy seems to make the demand for homogeneity, unity and what Lugones calls 'transparency' (as opposed to 'thickness') within the in-group even more pressing, and if this does not materialise, the success of the in-group in attacking the out-group is questioned.

Alternatively, thinking along Turner's (and, by implication, Lugones's lines) the protest itself could be understood as a suspension of existing hierarchical social structures (the existence of which should be acknowledged in the first place) and as a moment in which participants strive for universality among themselves (which is not necessarily the same as unity). One Indian ICC participant interviewed by Madsen (2001: 3733–4) pointed out that his coming to Europe had made him realise 'that not everybody in Europe is rich and happy', and that the situation of farmers in India and Europe is similar. From this experience, he concluded that 'the common people here and there were "made of common stock"'. Coming back to Lugones, 'sensing' resistance then becomes more about the intense encounter of the participants themselves, rather than the attack of the enemy. Fundamental structural critique and the envisioning of new possibilities emerge out of that encounter. In other words (and maybe not dissimilar to what happened in some of the Occupy and other more recent protest events), the structure to be critiqued does not pre-exist the protest (which is then organised against it, in order to articulate and transform it in a second step), but the structural critique (aware of multiple histories of domination and resistance) and the envisioning of new possibilities only emerges out of the encounter of multiple selves in the travelling of multiple 'worlds' that needs to come first.

Concrete body-to-body (street)walking

If the ICC is to be understood as a pilgrimage in a more literal sense than Lugones envisages, the concept of concrete space is crucial. In a section of the introduction to her book called 'Trespassing', Lugones herself (2003: 8) comments on 'the spatiality of resistances, within and against the spatiality of oppressions'. Her understanding, I argue, enables us to move beyond both the spatiality of the network and the one of fixity that some social movement theorists employ to understand the GJM past and present. Lugones asks her readers to visualise the workings of power as a map in which there is an allocated spot for each subjected subject. On that map '[a]ll roads and places are marked as places you may, must, or cannot occupy' (ibid.). Coming back to previously mentioned remarks about ICC participants, on such a map protesters would consequently be allowed to occupy sites of protest that are related to particular institutions and companies. They must not take any side streets or alleyways leading to sights they might want to visit as tourists, or those that might lead to business opportunities. For Lugones resistance to power needs to be able to break the abstraction of the map of power and to consequently allow for the concreteness of multiple movements of multiple, concrete selves. Only the abstract self has a spatiality that 'follows the designs of power' (ibid., 8) – concrete selves, through their concrete multiplicity, break the map qua definition and move according to a different 'relationality of space' (ibid., 11). While Deleuze and Guattari place the war machine in a *virtual* reality that cannot be grasped through the very categories of space and time, and that is resistant by precisely shining a light on how reality in its fullness transgresses the rigidity of what the actual allows, Lugones defines resistance as emerging from experiencing *actual* spatial and temporal reality as different in moving with others. This actual reality, in its concreteness, is different to what the logic of power wants us to believe it is.

Drawing on Michel de Certeau, Lugones (ibid., 211–12) describes those who draw the maps of power as strategists: they are the 'planners, managers, subjects of will and power' who abstract from the concrete 'from a point of view positioned high above the street'. By contrast, the concrete selves on the street are the 'weak', the ones that ' "insinuate" themselves into the other's place, fragmentarily, without taking it over in in its entirety, without being able to keep it at a distance' (ibid., 212). For de Certeau, the latter are not the *strategists* but the *tacticians*. Aiming to complicate this dichotomy, Lugones develops the concept of the 'streetwalker' who is a 'tactical strategist': similarly to what I have argued in the previous section about the primacy of the encounter, Lugones here defines the development of strategy as an act that emerges out of 'a long-winded intersubjective project' of concrete 'body-to-body engagement' (ibid., 207). Like the pilgrim in Turner's account, the streetwalker moves from 'hangout' to 'hangout', sharing space and time with others. However, in contrast to the pilgrim, the streetwalker more consciously aims to fundamentally disrupt and overcome existing structures. Developing 'a larger sense of the terrain and its social intricacies' through learning, listening, transmitting information and participation 'in communicative creations' allows her/him to sense 'directions of intentionality' on which s/he can build for her/his strategising (ibid., 209).

Many of the critical assessments of the ICC from the outside are implicitly based in the idea of protesters having to be straightforward strategists rather than streetwalkers in order to be able to adequately challenge the structures they critique. Though Stig Madsen, who has accompanied the ICC and on whose account I draw for much of my own analysis, shows appreciation and sympathy for the ICC as well as great insight into its internal problems and inconsistencies, he also (at least indirectly) holds the lack of overall strategy against it:

> I asked some of [the participants] why they protested against Nestle, and was told that Nestle was 'big' and had 'control'. . . . On the pavement a group of girls sceptically repeated the same question in their Swiss-French, but got no answer from the protestors who knew no French and next to nothing about Nestle. The Nanjundaswamy had not yet arrived; he would know the reason; he was the right person to answer the question.
>
> (Madsen, 2001: 3733)

At a seminar on the main square in Bern, Nanjundaswamy was asked: 'You only seem to have a destructive vision. What is your alternative?' He answered that, according to Marxist teachings, 'to make an omelette you have to break an egg'. As to how the Gandhian world [of the village republic] would look, he offered nothing. As the old Indian freedom fighter had said at the WTO demo: 'We do not want any trade organisation, international or national [. . .].' What does this vision of trade without trading associations boil down to in terms of alternatives? Caste, kinship and friendship as the basis of trade? . . . The truth, I think, is that the ICC has little to offer on how to embody ideas in institutions.

(ibid., 3735)

While the strategist could have answered the questions appropriately, having the right overview and knowledge, the non-knowledgeable participants were tacticians who, at their best, imposed themselves on a space they could not control and knew nothing about — a space they took over in its concreteness, but with insufficient abstraction. Madsen's account emerges from the vantage point of the seer who has, from the distance, an overview over what happens instead of being involved in body-to-body movement. It is unclear how Madsen understood his own role in relation to the ICC (maybe he understood himself rather traditionally as social observer and researcher, or alternatively as a sympathetic activist-researcher), but in this situation it is clear that he did not feel that it was up to him to answer the locals' questions that he refers to in the first quote above. Unintentionally his depiction here is rather close to the way the ICC was covered in the media, as for example this extract from a newspaper article demonstrates:

For many women, like Kumud Chowdhary of Gujarat, India, joining the caravan meant leaving their homes, family and village for the first time. They promptly packed pancakes, spices and pickles for their maiden voyage on an aeroplane . . . "I am here to 'Kill Monsanto' before it kills families like mine," quips Kumud, draped in her turquoise saree. She grows mustard and wheat and had not heard about genetically modified foods or the US corporation Monsanto until she became interested in the caravan.

(Bhandari, 1999)

Taking the article as a whole, it is unlikely that its author meant to deliberately belittle the ICC. Nonetheless, non-European participants are predominantly defined by a foregrounding of their lack of knowledge and ignorance of the overall context, plus a depiction of their protest in Europe as a 'freshman' one that is peppered with stereotypically colonial tropes: the 'maiden voyage' on the plane, the significance that is put on the difference in clothing and eating habits (the 'turquoise saree', the 'spices and pickles') etc. The boomerang perception of the (colonial) observer once again defines 'the other' by its difference to the neutral vantage point of the self (whose looks, eating habits or general life experiences are normally not deemed worthy of attention). For Lugones, by contrast, the tactical strategist gains insight into the social in a way that does not flatten it out, as it is not gained from a distant point of view. For the tactical strategist 'resisting ↔ oppressing has volume, intricacy, multiplicity of relationality and meaning, and it is approached with all the sensorial openness and keenness that permits resistant, liberatory, enduring, if dispersed, complexity of connection' (Lugones, 2003: 215).

What could have been the concrete differences in the organisation of and attitude toward the ICC had its participants been taken to be pilgrims and streetwalkers? Given that some of the non-European participants apparently wanted to see sights on top of protesting, organisers could have for example included a sightseeing tour that could have explicitly engaged with the local and global history of significant places. This tour could have taken the form of a participant-engaging encounter and education that could have gone beyond the immediacy of what the ICC was supposed to protest against, including for example an in-depth engagement with imperial and colonial European history, the history of the Holocaust in Germany etc. Such a tour would have deliberately complicated the confined and narrowly configured protest map, and would have paid tribute to the multiplicity of the participants' selves as well as the multiplicity of histories of domination and resistance. It might have led to encounters that could have complicated the structural critique and objectives of the ICC, potentially even leading the latter into a different direction. Such a tour might have also enabled participants to better deal with situations such as the one in which an Indian participant, when travelling through Germany, apparently praised Hitler as someone who 'defended the German nation state when it was in crisis, by getting rid of the problem elements', with India having to do the same (quoted in Anonymous, 1999: 29). Instead of attempting to understand the position of the participant, or even question how much this person could have been expected to know about a historical event that is arguably central to European history but not so much to those outside of Europe, the anonymous European ICC critic uses this example to question the ICC as such and to argue that this participant should have been immediately excluded (which, to her/his dismay, did not happen).

Drawing, among others, on the thought of Lugones, Jennifer McWeeny (2014) uses topographic maps as a heuristic device to advocate a feminist rethinking of connections among women. In contrast to other types of maps, topographic maps use so-called 'contour lines' to allow the reader to read a landscape three-dimensionally through the relationships between two-dimensional lines. Topographic maps include natural, cultural and historical features of a landscape, and are all about depicting the local (ibid., 277). For McWeeny (ibid., 278), feminist theory (which is what she is interested in) 'is at its best when, like a topographic map, it attends to the multiple dimensions of the spatiality of embodied experience and thinks its material, social, and historical aspects collectively and locally.' It implies a conscious looking out for what is hidden from one's own perspective, making 'visible the myriad perspectives . . . that constitute a world, despite our personal investments in concealing them.' Organising events that might have included for example visiting an exhibition on slavery or a Holocaust memorial in Germany might have led to a revealing of the multiplicity of selves and worlds that had previously lain hidden from the perspective and experience of participants on all sides, and for an encountering of difference outside of given frames. All of that might have been more challenging and disrupting of power than the organised protest at 'enemy' sites.

Connecting through 'things': becoming a faithful witness to oppression

So far this discussion has, in contrast to the earlier chapters of the book, purely focused on global environmental protest as a *human* phenomenon, with critique and reconceptualisation being targeted at *human* identity and practice. Given that a central objective of this book is the questioning of modern/colonial binaries, which are expressed in the human/nonhuman or culture/nature dichotomies, the question emerges how the thought of Lugones and the example of the ICC can contribute to that. In one of her interventions into decolonial thought Lugones herself (2010: 743) argues that 'the dichotomous hierarchy between the human and the non-human' is 'the central dichotomy of colonial modernity'. It is in opposition to the latter that she aims to develop 'non-modern' cosmologies, ecologies, economies and spiritualities (ibid., 742–3) that are able to do away with this and other dichotomies. As already pointed out in Chapter 1, the concept of the 'non-modern' is used in contrast to the concept of the 'pre-modern', which, Lugones says, is already subject to the modern, categorising logic (ibid., 743).

Lugones's emphasis on the significance of the nonhuman for the non-modern allows for possible links to Latour's thought and project. Interestingly, both Lugones and Latour use the term 'faithful' in (broad) relation to the significance of witnessing. For Lugones faithful witnessing emerges from an experience of the way that 'the manyness of worlds of sense' is suppressed, and it consequently 'conveys meaning against the oppressive grain' (Lugones, 2003: 7). That there are many worlds and multiple, complex sensing relates to a certain extent to Latour's claim that one can only be a faithful witness to the 'voice' of nonhumans if the latter are allowed to resist and transform the questions they are asked, which by definition questions the scientific experiment as a simple, 'clean' process. For Latour, nonhumans are not given proper recognition because of the dominant (modern) system of thought and practice (the two-house logic), though he shies away from calling this logic oppressive. For Lugones, the dominant logic denies the existence of the oppressed by subordinating it (and its capacity to express itself) to the (simple, pure, unified) existence of the oppressor. As I have shown in Chapter 2, Latour's political collective aims to provide an alternative to the dominant logic, while for Lugones alternatives can only be developed from within the resistance of the faithful witness to (colonial) oppression, in concrete, historical, 'impure' practices. To put it in Lugones's words (2010: 749), Latour provides a framework that is (rightly) 'fundamentally critical of the "categorical"/essentialist logic of modernity', but he does so without centring coloniality or the colonial difference.

Lugones questions the idea of 'common sense' as much as Latour, but for her the resistance of the faithful witness needs to take place 'in collision with common sense, with oppression', while Latour formulates an ideal scenario (the 'collective') that takes collision out of the picture.

Although Lugones states that overcoming the human-nonhuman dichotomy is central to developing non-modern thought, her own thinking contributes relatively little to how to get there. Indeed, her strong emphasis on the experience of the oppressed – who are *human* subjects – makes it occasionally difficult to properly think about how to move away from the centring of the human, and (by implication) from the modern/colonial project as such. In *Pilgrimages* nonhumans continue to be understood through the way they are experienced by the humans whose 'flesh and blood' forms the centre of Lugones's 'worlds' (Lugones, 2003: 87). How can nonhumans gain sufficient agency to actively shape 'worlds' outside of the experience and imagination of humans, and thereby fundamentally break with modern categories and concepts?

It is worth paying close attention to the way Lugones brings in the role and function of 'things' when weaving into her book stories of her own family and upbringing, particularly in relation to her mother. Indeed, when bringing up the notion of faithful witnessing for the first time, Lugones does so by recounting to what extent her mother had been a 'powerful resister' within a patriarchal system in which she could only identify herself as a housewife and mother. Lugones recalls the way her mother dealt with an event decades ago in which a vase that she had bought Lugones to give to a teacher as a present had broken, and which Lugones had consequently glued and then put on the shelf at home:

> "Well," [my mother] would say, "I wonder when and how that broke," as if it had just happened, rather than forty years ago. . . . After months of my asking her, she said, "I do it to remind myself that no one is listening." That was my mother, a powerful resister to any sort of cheap love anyone offered her, easy love by folks who were not ready to go the distance.
>
> (ibid., 8)

When Lugones speaks of the need to sense resistance, or to travel 'worlds' of sense, or to come to terms with the manyness of the 'worlds' of sense, her understanding of sense seems to be an intuitive one that does not require further definition. As previously mentioned, it is obviously meant to go against the notion of a 'common sense' that is monological and dominant. The significance of multiple sensing is grounded in an embodied, oppressed subjectivity that is, in its very existence, impure and complex. For Deleuze, sense 'is exactly the boundary between propositions and things' (Deleuze, 1990: 22). It points to features of things that are 'expressions' rather than 'possessions', and that make the substance of things become 'conceived as a self-positing power' (Strathausen, 2010). In other words, things *do* something; they express something rather than being describable in terms of their properties. The proposition then expresses something about things that does not 'pre-exist' in them but 'pre-inheres' (ibid.).

In Lugones's story about her mother, the resistance that Lugones is becoming aware of is inextricably related to the sense that is coming to the fore in the broken and then glued-together vase. It is what this vase *does* to her mother that allows for resistance to take place. In a related comment, Lugones (2003: 7) points out that her mother 'would sometimes – often – say things that were non-sense or false', like the comment about the vase ('I wonder when and how that broke'). The broken and then glued-together vase *makes* her mother formulate a proposition that is, following the 'common sense' of factual logic, 'non-sense', and it is only through that non-sense that faithful witnessing, as conveying meaning against the oppressive grain, takes place ('I do it to remind myself that no one is listening'). In a later chapter Lugones brings up 'arrogant perception' as a linchpin for the logic of oppression, and she points out that she herself, having grown up with the understanding that one can treat one's mother as an object to be (ab-) used, perceived her mother arrogantly (ibid., 78–9). Drawing on Marilyn Frye, Lugones defines arrogant perception as a way of perceiving others for oneself; as 'arrogat[ing] their substance to oneself' (ibid., 78). It is this 'arrogant perception' that has led Lugones, as a child, to perform a gesture of 'cheap love' in relation to the event with the vase, which her mother had well recognised and was not ready to accept.

In the story of the vase meaning is conveyed through sense beyond the true/ false distinction; indeed, it is conveyed through 'untruth' or 'non-sense'. The world that opens up and through which Lugones is finally able to travel (becoming a faithful witness to the oppression that it sheds light on) is determined not just by her mother's 'active subjectivity', but it is decisively shaped and determined by the nonhumans that are part of it. In Latour's sense, the vase does indeed *resist* the 'factual' framework of the common sense (which determines that it broke 40 years ago, was glued together by Lugones and then put on a shelf where it has been sitting ever since), by connecting to Lugones's mother (who becomes her faithful spokesperson) in order to express something else: 'I, the vase, remind your mother, and by implication you, that no one is interested in what she feels and thinks.'

Lugones tells another story about her mother, in which she states that it was to her father's 'advantage' (as 'part of his patriarchal position') that he was not able to 'follow [. . .] [his wife] in her moves' (ibid., 28), in contrast to her and her siblings who had to do precisely that and were therefore able to get her mother 'things' that the latter needed:

> Whenever my mother would ask for something, she would say "It is *on that thing next to that thing*." If you were not in the habit of following her in her moves – maybe that was not what your relation to her asked of you or what you put into it – *you would never be able to bring her "that thing."* . . . But if you did follow her into her moves, as we kids had to, you could easily get her "that thing." You see, she – someone who was to be unimportant, the perfection of whose makings was to lie in the making of not being visible – *managed to make herself important and to keep the makings both visible and*

invisible. "This," "on that," next to it," were stations her in path, she was the *pivotal directional subject.*

(ibid., 28–9; emphasis added)

It is once again 'things' that allow for Lugones's mother, in this instance, to become a powerful resister, a visible agent, a 'directional subject'. Following Lugones's broadly phenomenological approach, there can be no 'worlds' without subjects (who experience and inhabit it; cf. Ortega, 2001: 11), but these subjects do not (and potentially *cannot*) be understood outside of their connections to nonhumans. Instead of subjects narrowly conceived, one can centre human-nonhuman-assemblages as inhabiting a 'world' and bringing it about. Indeed, in this example it is once again 'things' that start the process of bringing to light both oppression and resistance: things in a spatial concreteness that allow for a different (streetwalking-enabling) relationality of space, against the abstractions of the map of power ('It is on that thing next to that thing').

What can be the consequences for a different understanding of the resistance of the ICC? I will turn to this in the conclusion of this chapter.

Conclusion: towards love and play in global (environmental) protest

For conventional political process theory, which dominates Anglo-American sociology, 'global movements' are nothing but extensions of older forms of social movements, with the process of globalisation creating 'new types of alliances between old types of actors'. Globalisation leads to a 'scale shift', which implies a shift of focus to international organisations 'that constitute a new international "political opportunity structure"' (McDonald, 2006: 19, quoting Jeff Goodwin and James Jasper). This approach is challenged by those scholars who argue, often drawing on poststructuralist theorists such as Foucault or Deleuze and Guattari, that contemporary global social movements do not bow to modern understandings of spatiality, identity and structure, but deploy a new kind of subjectivity as a weapon against the very logic that forms the basis of global governmental regimes. Enthusiasm about the emancipatory and disruptive potential of such new forms, understandings and objectives of political protest has early on led to the tendency to brush over structural hierarchies and oppression within parts of the GJM itself; which, however, have since been brought to light and critiqued by various critical social movement theorists.

As I have shown in this chapter, the ICC featured many elements of what has since the late 1990s been called the Global Justice Movement – both positively and negatively. What particularly came to the fore in the ICC was the difficulty that is involved in bringing together movements and activists from very different political and cultural contexts as well as historical backgrounds; movements that on the surface have the same concerns and visions, but that below that surface disagreed on at least as much as they agreed upon. As already mentioned in a previous section of this chapter, for Halvorsen (2012) contemporary grassroot movements such as the Occupy ones have for that reason been hesitant to immediately 'scale up' their grievances and critiques to the global level, and to straightaway connect them other movements across the world. Against the logic of networking and multiple, fluid connectivity, they have used, according to Halvorsen, the fixity of space as a weapon against the nature of contemporary capitalist systems.

However, from the perspective that I have taken in this chapter, the question should rather be what *kind* of connections social movements or activists should form, and how to go about this. Indeed, drawing on Lugones, connections that are formed in concrete body-to-body engagement are to be privileged to formulating strategies against an enemy from the distance. As I have shown, one of the central problems of the ICC was the lack of understanding the 'thickness' and multiplicity of particularly its non-European participants; leading to both positive and negative judgements that (from the outside and inside) quickly reverted to stereotypically colonial tropes, by depicting non-European protesters primarily as 'exotic yet ignorant' Others. For Lugones, by contrast, those resisting oppressions need to understand oppressions as intermeshed, taking place in different shades in different worlds that we need to be prepared to travel while letting go of the desire for transparency, simplicity, purity and sameness. Indeed, understanding both oppression and resistance as clear-cut and binary is inextricably bound to the vantage point of the 'strategist' who draws the map of power. This map, Lugones argues, works along the lines of the logic of the oppressor, by providing confined spaces for abstract, pure, simple, mono-sensical subjects. This is at odds with the corporeal, concrete existence of the oppressed, who experience the world by definition as 'world'-travellers that bring to the fore multiple histories of domination and resistance. The spatiality of oppressed subjects does not adhere with the logic of the map of power, but is located concretely on the street, testifying to a different relationality of space. Lugones calls for us to actively refashion ourselves in line with this understanding. We need to learn to *see* people intensively in all their facets and dimensions in order to move towards liberation.

It is from that perspective that I have attempted to develop a different perspective on the ICC and its liberating potential, one that regards ICC participants as pilgrims and streetwalkers who develop strategy out of the 'tactics' of concrete, unconditional, long-winded encounters with others. However, I have argued that a proper overcoming of colonial logic and perception also demands a shift of perspective to the role of nonhumans in this process of encounter; particularly in the light of Lugones's own emphasis on the need to overcome the human-nonhuman dichotomy that is so central to modern/colonial thought. Bringing together Latour and Lugones, I have maintained that the act of faithful witnessing against the grain of power does not only involve a witnessing of *human*, but also of *nonhuman* agency, which, following Latour, involves acknowledging the latter's capacity to challenge pre-given questions and frames of understanding.

I now want to conclude with outlining what acknowledging nonhuman agency and participation can concretely mean in the context of the ICC. In an earlier section of this chapter I have already referred to and critiqued Ainger's description of the ICC in her contribution to the book *We Are Everywhere* (2003). Now I want to return to the previously given quote, but this time from a different perspective:

> The following morning, I wake beside dozens of sleeping bodies, my hair filled with straw. . . . As I descend the wobbly ladder, the mist lifts off the surrounding fields, and I try to take in the improbable view below me. A rural Indian hamlet has been dropped wholesale into the German countryside. . . . A serene old man wearing kurta pajamas and plastic shoes in the centre of the courtyard takes a long, slow draw from a hookah pipe as large as himself, and still with its Aeroflot baggage-tag attached. He nods towards me graciously. In the far corner, steam rises from the vast vats of chai (tea) being brewed by local and international volunteers. . . . Women expertly wash themselves without removing their saris by the taps.
>
> (Ainger, 2003: 161)

As pointed out by Turner and already referred to in this chapter, the possibility of egalitarian communitas on a pilgrimage is mainly found in the 'pilgrim centers' where pilgrims eat, sleep and spend leisure time together. Communitas is furthered (or maybe even enabled) by 'things' such as food, drinks, beds, blankets, clothes, musical instruments, which facilitate the spatial concreteness that Lugones deems crucial for disrupting the abstract map of power.

The setting within which Ainger's observation takes place is very similar to the setting of a pilgrim centre: she describes (from her perspective) the scene that she found in front of her after she had woken up from communal sleeping, and the picture she takes in includes morning routines such as washing oneself and the preparation of breakfast. Though Ainger's description, as already pointed out earlier, is in many ways following a colonial logic, it still indicates the existence of the sort of communitas that Turner has in mind: a communitas that is about connections being formed between people that at least temporarily suspend social positions and given structures and create a community of equals. This communitas is one of concrete, multiple, 'thick' selves that travel multiple worlds. But what I want to add to this is the significance of the non-humans that contribute to, or even bring about, this connection in concrete, spatial terms: the straw that connects sleeping bodies, the lifting mists over the fields that bring to the fore the concrete spatiality of people, the connection between the baggage-tag, the pipe and the 'old man' that then (in his 'nodding'), again against the background of the lifting mist, forms a connection to Ainger herself. While one of the central problems of the colonial gaze is the lack of agency on the side of the colonised to actively shape and determine their (and other) 'world(s)', a shift of perspective to the agency of participating, connecting nonhumans allows both for the abandoning of a decisive element of the modern logic – the human-nonhuman dichotomy and its categorisation – and (relatedly) a multiple *sensing* of resistance beyond boomerang perception. It is their connection to nonhumans that gives ICC participants, in this instance, a spatial concreteness that goes beyond what is graspable from the point of view of the pure, distant and simple (colonial) observer. It is that connection that allows ICC participants to become pilgrims/streetwalkers who develop a sense of territory and relation with each other beyond the pre-given paths of the map of power designed by the strategists. But in order to fully become a faithful witness to both nonhumans and humans, and (by implication) to oppression, Ainger would have to let go of the 'beautiful soul' and become aware of the disruption and discomfort of difference and multiplicity (which, I would guess, she probably experienced on the ICC, but which gets no place in her description). Participating nonhumans do not just allow for 'feel-good' connections, but disrupt the sense of familiarity of the boomerang observer. What about, for example, the aching bones caused by sleeping in straw on a hard floor? What about unfamiliar and potential unpleasant smells? What about feeling cold, tired and hungry and (consequently) irritable? How can nonhumans, by causing these and similar sensations, nudge us to ask different questions that might challenge our frames of perception and experience?

For Lugones, arrogant perception has to be replaced with a loving one, and it is this love that can lead to faithful witnessing. Like Lugones had to do with her mother, activists might have to follow others' as well as their own connection to 'things' in order to fully understand the (multiple) workings of resisting ←→ oppressing, beyond and above the '(non-)sense' of whatever is written on protest banners, what is shouted (as slogans) on a demonstration, or the answers given in interviews to pre-given questions. Loving perception is linked to the need for 'playfulness', which is an attitude carrying an activity that is not determined by the essence and purpose of the activity (Lugones, 2003: 95). Playfulness, for Lugones (ibid., 96), 'is characterized by uncertainty, lack of self-importance, absence of rules or not taking rules as sacred, not worrying about competence, and lack of abandonment to a particular construction of oneself, others, and one's relation to them.' And it is, I would add, decisively enabled through the nonhumans that one engages in this activity, which transform selves and rules in the process of play.

In this chapter I have referred to many examples that testify to the failure to love and play on the journey of the ICC, both on the side of its participants and on the side of its observers. ICC participants became 'victims' of arrogant perception insofar that they were made 'pliable, foldable, file-awayable, classifiable' (ibid., 18). Similarly nonhumans simply became means to decorate a world that one already knew, as exemplified in the account of the journalist, but also to some extent in Ainger's depiction. But what Ainger's story at least indicates is the possibility of a perception that can take us beyond arrogant perception; allowing us to see its victims – including nonhumans! – as 'subjects, lively beings, resisters, constructors of visions' (ibid.). Focusing on the connecting and enabling agency of nonhumans provides one way to overcome arrogant perception and to bring about the 'active subjectivity' of those who resist, as faithful witnesses against the grain of power.

Notes

1 It is still controversial to what extent it was the protests that shut down the negotiations, or whether the main reason was the deadlock among participating states in the negotiations themselves.
2 I take the concept of 'thick' individuals from Lugones and will explain it at a later point in the chapter.
3 In this context the term 'third world' is used to describe the need for a breaking-apart of the binaries of human thought, not as a derogatory term that classifies various degrees of development (Sandoval, 1998: 355).
4 Lugones (2003: 16) deliberately keeps the word 'world(s)' in quotes, to draw attention to the heterogeneity and permeability of 'world(s)'.
5 Lugones draws on Elizabeth Spelman (who, she says, 'is remarkable among white feminist theorists in uncovering the many faces of feminist racism's resilience') to explain boomerang perception: 'I look at you and come right back to myself', says Spelman. 'In the United States white children like me got early training in boomerang perception when we were told by well-meaning white adults that Black people were just like us – never, however, that we were just like Blacks.' (Spelman quoted in Lugones, 2003: 157)
6 A note of caution is necessary at this point: my analysis here relies on selective statements and observations made by some ICC participants and accompaniers, so any generalisation is impossible.

Conclusion

> We need to stop seeing things as separate, solid objects with definite locations in space and time. Instead, we should see them as delocalized, indefinite, mutually entangled entities.
>
> Mae-Wan Ho (2000)

> It is simply the colonial difference that is at stake . . . a room looks altered if you enter it from a different door . . . of the many doors through which one could have entered the room of philosophy, only one was open. The rest were closed.
>
> Walter D. Mignolo (2002)

Science-based anti-GMO activists reject the claim that the gene is a separate, self-contained entity that determines the development of the organism. Accordingly, they also dismiss the idea that the effects of gene manipulation are predictable and controllable. As this book has shown, at stake in this argument are the central dichotomies of modern thought: subject/object, mind/matter, man/nature. Drawing on complexity science, critical anti-GMO scientists such as above-quoted Mae-Wan Ho argue for an understanding of organic development that regards it as interrelated and indefinite, with agency being diffused across the different parts of the organism. This position is closely related to the attempt of critical scholars such as Latour and Deleuze to develop new, non-binary ontologies (or even metaphysics, in Deleuze's case) of matter, materiality, agency, nature, transformation and politics. However, this book has maintained that any attempt to overcome the binary thinking of modern thought and practice (including the mainstream scientific practices and theories to which it has given birth, grounding among other things GE technology) first needs to come to terms with how the rise of such thinking has gone along with the oppression and destruction of other ways to know and live. While most critical scholars and activists would readily acknowledge the existence and significance of that oppression and destruction, its deeper epistemological and ontological dimensions and implications – the 'coloniality of knowledge' (Mignolo, 2007: 451) and the 'coloniality of Being' (Maldonado-Torres, 2007) – too often remain unchallenged. Mignolo's point about the 'closed doors to the room of philosophy' that is cited above (Mignolo, 2002: 65) is not brought to bear upon theorisation: the latter continues to take place from the vantage point and privilege of those who have always been able to access the room of philosophy through the one open door, with given (European) concepts at hand. Alternative ontologies of abundance and interconnection are too often developed without systematically making sense of the fundamental *dis*connection that has co-constituted the rise of modern/colonial binary thought and practice. This is not only a problem for philosophy and theory-development, but also for the practices of knowledge production among political activists, whose analysis of domination and strategies to undo the latter often ignores the nature of colonial violation. As decolonial scholars have pointed out, colonial practice has not just shaped actual structures of oppression and exclusion, but has produced (and continues to reproduce) the very categories we have at hand to make sense of such structures (cf. Coleman, 2015a 2015b; Coleman and Rosenow, 2016; Coleman and Rosenow, 2017a, 2017b; Ansems de Vries et al., 2017).

The aim of this book has been to think about how we can support the desire of anti-GMO activists to advance different, non-binary understandings of nature and the world and thereby critique modern ways of thinking and acting more generally, while at the same time overcoming the problem of colonial difference. Writing the book has been a project of transition in the process by which I myself have come late to realising how no analysis of nature, environmental activism, political theory and philosophy can proceed without reference to the significance of coloniality. As a result of this trajectory, the book clearly features important omissions, most notably the lack of an actual encounter with the people whose experiences of oppression I am trying to delineate and make sense of (a lack that by definition leads to the danger of ever-new violation and patronisation), as well as the lack of thorough engagement with non-modern cosmologies and metaphysics in relation to nature and GE technology (cf. Todd, 2016). However, a book in transition that is clear about these omissions and problems is still better than a book that would have been fully coherent on its own terrain, but that had omitted any reflection on coloniality.

This conclusion is meant to give a more elaborate outline of the position that I have reached in this book: pointing at and connecting key themes and ideas. Because it has been a work of transition, not all themes and ideas have been there from the beginning, and the overall approach has emerged in the process of writing. The conclusion shows how a decolonial position (that is also informed by postcolonial argument) allows for a way of breaking free from the tensions and exclusions that pervade binary modern/colonial forms of thought and practice, and the implications that such thought and practice has for thinking about and acting in relation to nature, society and political transformation. The conclusion will highlight once again both the potential as well as the limits of existing anti-GMO arguments and practices in relation to the conceptual resources that I have used. It will also present a retrospective reflection on the sort of methodological approach that I have (implicitly) developed throughout, to be taken further in my work to come. Last but not least, given that this book has also aimed to make a concrete intervention that is relevant for anti-GMO activists, the conclusion will end with a seven-point activist 'manifesto'.

Towards (more) reality

My engagement began with an analysis of the arguments brought forward by those GMO opponents who draw on science to make their critique. Developmentalist biologists, for example, critique the neo-Darwinist Central Dogma of mainstream molecular biology, which (to use María Lugones's term) *split-separates* organic development into the clear-cut parts of gene → protein → organism. As I have argued in Chapter 2, drawing mainly upon Latour, invoking science to advocate a different politics towards nature is not per se problematic. For Latour the problem is rather the radical constructivist approach towards science, which in its rejection of reality continues to embrace what he calls the modern two-house logic of 'nature' as opposed to 'the social'. Constructivists, Latour maintains, affirm this logic by leaving the study of 'nature' to the scientists alone, arguing for a politics that is solely based in social processes and understandings.

However, embracing the existence of *reality* is not the same as embracing the concept of *facts*. Facts continue to be grounded in the two-house logic insofar as the house of nature is determined as the location of facts, and the house of the social as the location of values and beliefs. Following this logic, scientists, as 'experts', are the only ones able to travel from one world to the other to inform society about objective natural laws. Facts serve to close down discussions and reduce the agency of the nonhuman world to the one of an analysable object. In order to overcome the two-house logic altogether, Latour maintains that we need to move towards an understanding of reality that depicts it as countless, ever-changing, dynamic human-nonhuman associations. Those associations make propositions to what Latour calls 'the collective' (a term that is supposed to replace both nature and society). Whether these propositions are to be included into the collective needs to be decided by a slow process of deliberation. Propositions, in contrast to facts, do not close down discussions, but rather kick them off. They are unforeseen and puzzling, leading to 'uncertainty and not arrogance' (Latour, 2004: 83).

Returning to the question of anti-GMO argument, if the GMO becomes part of such a scenario, it can no longer simply be regarded as an unnatural, artificial entity the monstrosity of which is a given 'fact'. Instead it becomes part of a human-nonhuman association that has the right of making a claim for inclusion to the collective. This claim should be deliberated without recourse to the notion of fact, but there are still grounds for excluding it, depending on what the deliberation of the collective takes into account. As I have shown in Chapter 2, if the collective is no longer taken to be an ideal scenario, but an actual (to use again Lugones's terminology) *impure* body, such deliberation might well take into account the multiple histories and experiences of oppression in the light of which a plea for inclusion is made; without having to refer to the 'facts' of nature and science. For example, I have maintained in Chapter 2 that an exclusion of the neo-Darwinistic proposition 'survival machine', which depicts the organism as an egoistic entity that fights for survival in a hostile environment, and which is based on the same theoretical ground as the technology of GE – the Central

Dogma – can be justified by arguing that its underlying logic is colonial. The logic of the proposition 'survival machine' is the same that Sylvia Wynter identifies as the logic of white overrepresented Man (and which I have, due to the congruency, consequently termed the logic of 'Man/gene'). Excluding the GMO from the collective becomes, in this scenario, a political act that is able to take into account the links to colonial history. In this history overrepresented (white) Man started to fully saturate the idea of the Human, with every difference being one of aberration. Similarly, the gene has been regarded as the full and only agent of organic development.

Most anti-GMO activists dispute the scientific consensus that GMOs are safe to produce and consume, but many do so from a position that embraces the idea that science is still *the* privileged site for producing factual knowledge about a nature that is somewhere 'out there'. Pointing towards complexity, uncertainty and unpredictability is central to much anti-GMO argument, but paradoxically the 'nature' of the GMO remains beyond doubt: it is *predictably* unnatural, monstrous and damaging, and it (consequently) has no right to exist. As I have pointed out in Chapter 2 with reference to Ho's argument, 'nature' remains an uncontaminated space that should be a model for the organisation of society – its truth existing independently of history and power, though the latter has, it is argued, contributed to nature's deterioration. Indeed, 'facts' and certainties that anti-GMO activists refer to and hold on to do not just emerge from the site of science, but also from their understanding of power and domination: Monsanto, it is argued, is a key driver of neoliberal globalisation, with GE technology being a means of oppression and exploitation of the poor in the Global South as well as of nature, with a return to small-scale (organic) agriculture as the only sustainable alternative. As I have shown in Chapter 3, referring to such 'facts' of the situation is in danger of overlooking the particularities of the colonial and postcolonial trajectory in a place such as India, and of confining the agency of both subaltern humans and nonhumans within the very categories that continue to reproduce colonial violation. What I have called, in Chapter 3, the logic of Man/gene/State continues to be a dominant force in contemporary Indian agricultural politics, and it grounds the ongoing strong belief in science, progress and political economy that has been decisive for the founding of the Indian postcolonial state, often *in continuity* with colonial paradigms. In other words, the problem is not just external forces such as 'the West' and multinational corporations, or even the internal neoliberal course of the Indian state since the 1990s, but the identities, epistemes and political as well as 'natural' materialities that have been shaped by such forces for centuries, and that have been central for making what Deleuze calls the 'capitalist axiomatic' work in this context.

Following Latour, affirming science as a decisive means through which non-humans can (in association with humans) make their presence be felt is not the same as holding on to the notion of facts. If science is understood as giving speech prostheses to ever-new human-nonhuman associations, and to thereby cause perplexity, surprise and discomfort, it can be *one* useful way of challenging dominant ways of understanding the world (as for example seen with the rise of complexity theory; cf. Chapter 2). However, importantly it is not the *only* way. As I have shown in Chapter 3 by drawing on the accounts of anthropologists Gregg Hetherington, Susana Carro-Ripalda and Marta Astier, there are other, *sensual* ways about which I will talk in more detail in one of the following sections. All of these accounts point to the need to understand reality as definite and non-definite at the same time, involving the participation of both humans and nonhumans, and making associations between matter, history and power in particular configurations. Anti-GMO activists have to become open to the surprise and perplexity of that reality, in order to have their dominant frames of understanding challenged and transformed, and to address the problem of colonial violation.

Reflections on method

Unfortunately, Latour's *Politics of Nature* features no proper engagement with the colonised Other of the two-house logic. Correspondingly, his suggestions about an alternative ontology of the world and an alternative organisation of politics within the collective makes only few, non-specific references to other knowledges and cosmologies that have challenged the two-house logic long before it had become a matter of concern for Latour (cf. Todd, 2016). To go once again back to the opening quote by Mignolo, Latour went about his business of philosophising in the room of philosophy without much reflection (beyond a mere stating) on why there are so many closed doors, and/or what might lie behind those doors. He continues to remain guilty of what Enrique Dussel has once called, in relation to the thought of Gianni Vattimo, 'a eurocentric critique of modernity' (quoted in Mignolo, 2002: 57).

On the other hand, the danger of some decolonial arguments is that their call to reckon with the *historical reality* of modern/colonial co-constitution too easily follows an approach to history and reality that is once again deeply modern. The historical trajectory and account of global development that we find in, for example, Mignolo, Quijano and Escobar leaves little space for a narrative from within the non-modern cosmologies that they aim to foreground.[1] Indeed, the latter cosmologies might feature different temporalities and spatialities; they might refer to spirits, ancestors and 'things' as sentient beings and decisive agents. Here, Latour's call for acknowledging the presence of nonhumans by providing space for them to resist and transform the questions we ask of them might be, in substance, closer to some of the cosmologies that have been annihilated by the rise of modernity than the frameworks of decolonial scholars themselves. Moreover, Latour's acknowledgement of the reality of nonhuman existence is able to provide a ground for the challenge of his *own* frame (though this is probably not intended). It is unclear to what extent a similar challenge to taken-for-granted frames can be launched from within the thought of prominent decolonial scholars (for a wonderful exception, see Suárez-Krabbe, 2016).

It can be argued that what needs to be pursued (and what I have pursued, to some extent, in this book) is a non-systematic dialectical approach. Understanding and coming to terms with the implications of colonial violation, and consequently critiquing modern thought and (binary!) categorisation is in some ways related to following a traditional, modern approach to questions of history and reality. But that understanding should then lead to a priorisation of knowledges and cosmologies that come from what Mignolo calls the 'outside',[2] and which might challenge that very understanding of history and reality. 'Outside' ontologies might feature understandings of nonhuman agency and matter that lead to some crossovers and fruitful dialogue with European critics of modernity such as Latour and Deleuze. Of course it can be argued that such an approach is in danger of yet again reproducing colonial violence, by making 'outside' ontologies 'legitimate' through reading them through, or in relation to, 'legitimate' Western scholars (cf. Tilley, 2017). In this book I have tried to get away from this problem by subordinating the Western thinkers that I have engaged to their function of shedding light on the existence of what I have called, drawing on Carro-Ripalda and Astier (2014), 'ontological incompatibility', particularly in Chapters 3 and 4.[3] Moreover, from a potentially more *post-* than *decolonial* perspective I would also maintain that the move of bringing European critics of modernity close to non-modern ontologies contributes to challenging the idea that the latter are mainly characterised by their exotic 'difference' to what 'we' – Europeans – know.

Sense and love: beyond the monologue

While for Latour reality is made of ever-changing conglomerates of human-nonhuman association, Deleuze formulates a metaphysical conception of reality that configures the field of the 'actual' as only one of two. The actual is complemented by the virtual, which is different in kind. The virtual is prior to the actual, and is made of non-identical, non-comparable, multiple intensive forces. Similarly to Lugones, for Deleuze multiplicity does not necessarily have to lead to unity, and the multiplicity of the virtual is indeed of a kind that cannot be unified. The virtual cannot be accessed through the categories that we have at hand – representation, deliberation, rationalisation etc. – but it can be *sensed*, and it is precisely through such *sensing* that our taken-for-granted identities can open up to the experience of difference in itself. As I have pointed out in Chapter 4, for Deleuze, sense is not a faculty of the subject that is exercised in relation to an object. Instead it is what lies between 'propositions' and 'things'. Sensing undoes the subject by no longer making it define itself by its substance, but by connecting in various, multiple ways to other 'things' and realities – realities in which things are defined by their 'self-positing power' (Strathausen, 2010). What is sensed is not what has pre-existed prior to the sensing, but what has 'pre-inhered' (ibid.).

Lugones also uses of the concept of sense, though always in relation to experiences that are lived. Being able to *sense* means, for her, that we need to escape the logic of domination, which advances the idea of the self as pure, simple, distant and coherent. Instead we need to actively shape our subjectivity in line with the reality of the oppressed, in which selves exist as multiple, as travelling 'worlds', as finding themselves in the midst of multiple histories of oppression and resistance that are not always commensurable. Reality is multiple, existing as multiple 'worlds'. But in contrast to Deleuze, Lugones determines all of these 'worlds' as being inhabited by flesh and blood people (though imagined or dead people can also inhabit them in connection to the living). For Lugones, experiencing or shaping an 'active subjectivity' allows for a *sensing* of resistance, beyond the monological and mono-sensical logic of the oppressor. In ways not dissimilar to Deleuze, experiencing the 'manyness of worlds of sense', which structures of power fragment and keep apart (Lugones, 2003: 6), undoes the subject. It does so not by connecting it to a reality that is virtual, but by connecting it to the experience of those oppressed subjects who have always experienced themselves as multiple, complex and incoherent.

An intuitive notion of sense is also embraced by many anti-GMO activists, particularly in their accounts of nature as beautiful, dancing, song-like, colourful, vibrant and harmonious. But the problem of many of these accounts is that they only sensually grasp of nature what they already think they know about it, confirming 'feel good' sensations that subordinate all experience to what Deleuze (2004: 64), drawing on Hegel, calls the 'beautiful soul'. The beautiful soul is only able to see a difference insofar that it is federative, reconcilable, harmonious. The logic of the beautiful soul resembles Lugones's (2003: 157) critique of white people's boomerang thinking (for which she draws on Elizabeth Spelman): difference only becomes recognisable in relation to the norm that one finds in oneself, emerging from and returning back to oneself. As Chapter 4 has shown, the logic of the beautiful soul underlies not only the mainstream environmental activist's understanding of nature, but also her/his understanding of the exotic Other. The latter's difference is accommodated if it fits one's given frame, but rejected if it throws the latter into doubt.

The notion of sense that comes to the fore in both Lugones and Deleuze (despite their philosophical differences) does not cause comfort, but bewilders and puzzles. Instead of fitting everything together, it undoes. It leads beyond the ability to categorise, and it fundamentally disrupts binary logic. Following Deleuze, it allows for sensing the Being of the statement beyond the said, and the Being of the visibilities beyond the seen. However, as I have pointed out in Chapter 3 in relation to the attitude of Indian farmers in relation to Bt cotton, this sensed Being is not necessarily one of *more*, but also of *lack* – leading to the need to endure the silence that has resulted from the eradication of alternative visions and cosmologies. As Chapter 4 has shown, instead of grappling with and attempting to endure this silence and the discomfort it causes, activists often turn to making accusations about a lack of participants' authenticity, for example by saying that those who hold Green Revolution-affirming positions are not 'true' peasants. Or alternatively, by depicting those who hold such positions as victims of manipulation, as we have seen in the case of Shiva (cf. Chapter 3).

Sense is a concept that is lacking in Latour's analysis. One can argue that Deleuze 'practices' his own metaphysics by employing a particular style of writing, which has led commentator James Williams (2005: 15) to maintain that part of Deleuze's success lies in how he is making her/his reader *feel* that he is right. Relatedly, Lugones believes in theorising only insofar that it emerges out of lived practice, and for her this 'requires a metamorphosis of self in relation as well as a metamorphosis of relations in defiance of both individualism and privacy as the domain of one's affective belonging' (Lugones, 2003: ix). By contrast, in *Politics of Nature* Latour seems overly detached from his own collective, which, para-doxically, is predominantly characterised by attachment and connectivity (e.g. Latour, 2004: 21). While New Materialist scholars such as William Connolly or Jane Bennett make sense of materiality by drawing at least to some extent on their own encounters, and by outlining their own sensational and political attach-ments beyond any rational schemata, Latour's 'middle' work comes across as formalised, schematic, technical and driven towards institutionalised order. At the same time, one of the decisive questions that Latour wants the collective to ask – 'How many are we?' – is supposed to go against feelings of self-importance, arro-gance and fixed coherence – the latter being all features that Lugones defines as belonging to the oppressor's logic of 'arrogant perception'. 'How many are we?' is a question that can only be posed *and* answered from the position of a multiple, multi-sensual, resisting self. Arrogant perception is unable to make sense of new, *other*, unforeseen human-nonhuman associations and propositions, because it takes place from the vantage point of the oppressor who sees the landscape below, from the distance, as abstract and fixed.

Making sense of the world beyond the logic of common sense, based on what Lugones calls *loving* perception, allows both nonhumans *and* oppressed humans to challenge given frames and categories of understanding, and thereby transform a world that has been built on their exclusion and ontological eradication. As I have outlined in Chapter 4, this means that the development of strategies for tack-ling oppression needs to be subordinated to what arises in the unconditional, con-crete, sensual encounters with others. Anti-GMO activists (and political activists more generally) need to *sensually* encounter others (both human and nonhuman, or maybe, as outlined in Chapter 4, the human *through* the nonhuman) in order to *sense* resistance and overcome what Lugones calls enmeshed oppressions. This involves a suspension of the importance of the self, and often, going along with that, a suspension of the focus on the final objective, on the injustice one wants to do away with, and the clarity that one thinks one has about it.

Streetwalking: developing strategies out of concrete encounters

As I have shown in Chapter 4, Lugones's understanding of the spatiality of con-crete selves as going in their very existence against the abstract maps of power designed by distant strategists complicate the spatial and temporal imaginaries of global protest. For Lugones, the right question is not whether social move-ments should reach out to and connect to 'the global' or not, but what *kind* of connectivity we should nurture. Drawing on Lugones and Viktor Turner's notion of the pilgrimage, I have argued in Chapter 4 that political activists should sup-port a connectivity that draws on the spatiality and temporality of the *pilgrim* or the *streetwalker*. For Turner, pilgrims strive for what he calls *communitas*, which means an equality between participants that temporarily (i.e. on the pilgrimage) removes differentiated social positions and structural inequality; or that at least, as

Turner (1974: 207) puts it, takes the 'sting' out of existing social hierarchies. This is decisively enabled through sharing food, places to sleep and leisure in pilgrimage centres, while the actual sites of worship are mostly located in hierarchically-structured social spaces. Turner defines the striving for *communitas* among the pilgrims as being different from the striving for *solidarity*: while the former nurtures universality within an in-group, the latter forms the in-group primarily in relation to an out-group (an enemy). As I have argued in Chapter 4, protest movements like the Intercontinental Caravan are focused on solidarity instead of communitas: they define their commonality as what unites them against the structures of oppression that they aim to target. Focusing on the out-group easily leads to dismissal, judgement and arrogant boomerang perception if participants of what is perceived as the in-group do not fit the frame one requires of them. Because everyone is focused on attacking the enemy, there is little acceptance of internal complexity, of letting one's own strategies of sense-making be challenged, or of disrupting the in/out binary in general. I have suggested that political activists should instead understand themselves as pilgrims who encounter others in concrete spaces of togetherness and who develop general strategies as well as understandings of oppression out of these encounters. Only then are they able to take into account multiple histories of oppression, multiple selves and develop an awareness of how what makes sense in *one* world does not necessarily do so in *another*. Lugones defines as *streetwalkers* those who acknowledge the significance of developing strategies out of the concreteness of encounters on the street ('hanging out'), rather than from the distant vantage point of the seer who can only take in the street below as abstract.

Environmental problems by definition transgress bordered human communities.[4] Concrete connections need to be formed among those who want to stand up for protecting nonhuman realities, but the focus on protecting 'nature' must not divert us from the need to reflect on the impact of (colonial) oppression on human as well as nonhuman realities and experiences (cf. Braun, 2002). What problems we face in relation to the 'environment', how we should go about them in political protest, how we should frame them, and what should be done about them – all of these are questions the answers to which should be up for grabs. The answers (and maybe even the questions!) should not precede connections and encounters, but emerge out of them. In these encounters, we need to actively shape our subjectivities to be the ones of 'world'-travellers; travellers who are acutely aware (following Spivak) of their own geopolitical positioning and potential complicity in reinforcing structures of oppression through advancing dominant frames. As I have shown in Chapter 4, nurturing a different understanding of the self and of oppression needs to make space for the agency of nonhumans in bringing about what Lugones calls a non-arrogant, loving perception. If we follow our *sensing* of nonhuman presence, we are able to move beyond the colonial gaze, and open ourselves to experiences of difference that lead to uncertainty, perplexity and disturbance. Focusing on our own subjectivity and on the encounter with others can be frustrating, given the urgency of some of the problems that we face in today's world. Nurturing different understandings of the self, of the world around us, and of how to do politics, is a painfully *slow* process, as both Latour and Lugones are keen to emphasise. Such approach does not provide or advocate quick solutions, and it thereby truly goes against what Lugones (2003: 7) defines as 'the grain of power'. Though it is frustratingly slow and complex, it might be, as Spivak points out, the only way to not be complicit in the constant reproduction of oppression.

An anti-GMO activist manifesto

(1) Science, but not facts

We need to affirm the significance of science, and we need to continue fighting Man/gene/State by questioning the conventional scientific paradigms on the basis of science itself. But we need to broaden our understanding of what counts as legitimate (and even as scientific) knowledge by engaging with the ways knowledge about 'nature' has been accumulated and experimented with by non-modern peoples. Empirical science, in a broad sense, can be a means for nonhumans to challenge our frames and to make their presence felt. But there are also other ways for that, with which we have to engage. We have to get away from the notion of fact and start to analyse those propositions that emerge out of complex, specific, human-nonhuman encounters in concrete (and often different) contexts, and we need to deliberate those against the background of centuries of colonial violation.

(2) GMOs have a right to exist

We need to stop understanding GMOs as unnatural, monstrous entities the identity of which we fully know and reject. GMOs, and the particular human-nonhuman association they are part of, have a right to ontological self-definition. If nature is unpredictable, complex and mysterious, so are GMOs. That does not mean that we should include them in the collective. But if they are to be excluded, this might take place on other grounds than their capacity to cause harm to health and environment. A better ground for exclusion might be a reference to the colonial logic that underlies the scientific premises that have brought them about.

(3) Nature is not beautiful

We need to stop understanding Nature as ahistorical and scientifically knowable. We need to stop perceiving Nature as One. We need to de-unify and de-beautify. We need to start regarding nonhumans as forces that can and should challenge our given frames of the world: causing uncertainty and perplexity instead of feel-good sensations of harmony. We need to stop labelling certain nonhumans (as well as humans!) 'unnatural' or 'natural' (or 'traditional' or 'authentic').

(4) Pro uncertainty

We need to embrace uncertainty, perplexity and disturbance, as it can undo the logic of oppression by undoing coherence (of subjects, of problems, of solutions).

We need to stop asking pre-given questions that make sense in our pre-given frames.

(5) No to strategies, yes to tactics

We need to become streetwalkers instead of strategists. We need to develop our bigger strategies out of concrete, long-winded, embodied movements with others. We need to appreciate the 'weak', the ones employing tactics, the ones ignorant of strategy, the ones who are taking over the concreteness of the street. We should not belittle them or accuse them of their lack of knowledge or their lack of willingness to abstract.

(6) Listen to the oppressed – and act accordingly

When it comes to understanding the greatest problems of the world (including environmental degradation), it is the experience of the oppressed and their understanding of the world that needs to be prioritised. To paraphrase Russell Means (1980): we can only judge the theories about how to solve the world's problems 'by the effects it will have on non-European peoples' (or, I would add, on those with non-European/non-Western roots). This is because every revolution in European history has served to reinforce Europe's tendencies and abilities to export destruction to other peoples, other cultures and nonhuman existence.

(7) Be slow

We need to resist the temptation of urgency, we need to resist the temptation to short-circuit in order to 'solve' environmental problems. Only slow deliberation and intersubjective encounter can avoid further oppression.

Notes

1 As already explained in chapter one, I draw on Lugones (2010) for the term 'non-modern'.
2 As already explained in Chapter 3, similarly to the concept of the 'non-modern', the concept of the 'outside' does not refer to an 'ontological outside', but to 'an outside that is precisely constituted as difference by a hegemonic discourse' (Escobar, 2007: 186).
3 I have probably done this more faithfully in relation to Deleuze than to Latour. This is once again due to the transitional nature of this book. Latour has been mainly engaged in Chapter 2, which is still closest to the original manuscript.
4 It needs to be noted that this is also related to a particular discursive frame: as Christopher Rootes notes with reference to the British context, the global framing of environmental issues since the mid-2000s, particularly in relation to climate change, has made more local issue-centred environmental NGOs, which might have previously focused on wildlife protection or the reduction of waste, to change their attention to 'universalizing' discourses and concerns (Rootes, 2012: 18).

References

Abrahamsson, S., Bertoni, F. and Mol, A. (2015) Living with omega-3: New materialism and enduring concerns. *Environment and Planning D* 33(1): 4–19.

Ainger, K. (2003) Life is not business: The intercontinental caravan. In: Notes from Nowhere (ed) *We Are Everywhere: The Irresistible Rise of Global Anticapitalism*. London: Verso, 160–70.

Alessandrini, D. (2010) GMOs and the crisis of objectivity: Nature, science, and the challenge of uncertainty. *Social & Legal Studies* 19(1): 3–23.

Anonymous (1999) The intercontinental caravan: A critical analysis. *Do or Die* 8: 28–9.

Ansems de Vries, L., Coleman, L.M., Rosenow, D., Tazzioli, M. and Vázquez, R. (2017) Fracturing politics (or, how to avoid the tacit reproduction of modern/colonial ontologies in critical thought). *International Political Sociology* 11(1): 90–108.

Ansems de Vries, L. and Rosenow, D. (2015) Opposing the opposition? Binarity and complexity in political resistance. *Environment and Planning D* 33(6): 1118–34.

Anzaldúa, G. (1987) *Borderlands/La Frontera: The New Mestiza*. San Francisco: Aunt Lute Books.

Atkinson, P., Glasner, P. and Lock, M. (2009) Genetics and society: Perspectives from the twenty-first century. In: Atkinson, P., Glasner, P. and Lock, M. (eds) *Handbook of Genetics and Society: Mapping the New Genomic Era*. London and New York: Routledge, 1–14.

Barry, A. (2013) *Material Politics: Disputes along the Pipeline*. Chichester: Wiley-Blackwell.

Barry, A. (2005) The constituents of Europe: Speech on the occasion of the finissage of the exhibition *Making Things Public*, 2 October. Available at: http://zkm.de/en/andrew-barry-the-constituents-of-europe

Bhambra, G.K. (2014) *Connected Sociologies*. London et al.: Bloomsbury Publishing.

Bhandari, N. (1999) South Asian farmers take protest to London. *IPS Press Release*, 30 May. Available at: http://caravan.squat.net/articles/ICCframe-art.htm

Blaser, M. (2013) Ontological conflicts and the stories of peoples in spite of Europe. *Current Anthropology* 54(5): 547–68.

Braun, B. (2015) New materialisms and neoliberal natures: The 2013 antipode RGS-IBG lecture. *Antipode* 47(1): 1–14.

Braun, B. (2002) *The Intemperate Rainforest: Nature, Culture and Power on Canada's West Coast*. Minneapolis and London: University of Minnesota Press.

Braun, B. and Whatmore, S. (2010) The stuff of politics: An introduction. In: Braun, B. and Whatmore, S. (eds) *Political Matter: Technoscience, Democracy, and Public Life*. Minneapolis and London: University of Minnesota Press, ix–xv.

Calvert, J. (2007) Patenting genomic objects. *Science as Culture* 16(2): 207–23.

Carolan, M.S. (2008) From patent law to regulation: The ontological gerrymandering of biotechnology. *Environmental Politics* 17(5): 749–65.

Carro-Ripalda, S. and Astier, M. (2014) Silenced voices, vital arguments: Smallholder farmers in the Mexican GM maize controversy. *Agriculture and Human Values* 31(4): 655–63.

Carro-Ripalda, S., Astier, M. and Artía, P. (2015) An analysis of the GM crop debate in Mexico. In: Macnaghten, P., Egorova, Y. and Mantuong, K. (eds) *Governing Agricultural Sustainability: Global Lessons from GM Crops*. London and New York: Routledge, 33–73.

Caruso, D. (2007) Change to gene theory raises new challenges for biotech. *The New York Times*, 3 July. Available at: www.nytimes.com/2007/07/03/business/worldbusiness/03iht-biotech.4.6471136.html?pagewanted=1

Castelao-Lawless, T. (1995) Phenomenotechnique in historical perspective: Its origins and implications for philosophy of science. *Philosophy of Science* 62(1): 44–59.

Cerier, S. (2016) GMO phobic? The real Frankenfoods might surprise you. *Genetic Literacy Project*, 14 August. Available at: www.geneticliteracyproject.org/2016/08/14/gmo-phobic-real-frankenfoods-might-surprise/

Chandler, D. (2012) Development as freedom? From colonialism to countering climate change. *Development Dialogue* 58(April): 115–29.

Chandler, D. and Reid, J. (forthcoming) 'Being in Being': Contesting the ontopolitics of indigeneity today. *The European Legacy*.

Chesters, G. and Welsh, I. (2006) *Complexity and Social Movements: Multitudes at the Edge of Chaos*. London and New York: Routledge.

Chesters, G. and Welsh, I. (2005) Complexity and social movement(s): Process and emergence in planetary action systems. *Theory, Culture & Society* 22(5): 187–211.

Chow, R. (2012) *Entanglements, or Transmedial Thinking about Capture*. Durham and London: Duke University Press.

Colebrook, C. (2002) *Gilles Deleuze*. London and New York: Routledge.

Coleman, L.M. (2015a) Ethnography, commitment and critique: Departing from activist scholarship. *International Political Sociology* 9(3): 1060–75.

Coleman, L.M. (2015b) Struggles, over rights: Humanism, ethical dispossession and resistance. *Third World Quarterly* 36(6): 1060–75.

Coleman, L.M. and Hughes, H. (2015) Distance. In: Aradau, C., Huysmans, J., Neal, A. and Voelkner, N. (eds) *Critical Security Methods: New Framework for Analysis*. London and New York: Routledge, 140–58.

Coleman, L.M. and Rosenow, D. (2017a) Beyond biopolitics: Struggles over nature. In: Prozorov, S. and Rentea, S. (eds) *The Routledge Handbook of Biopolitics*. London and New York: Routledge, 260–70.

Coleman, L.M. and Rosenow, D. (2017b) Mobilisation. In: Bilgin, P. and Guillaume, X. (eds) *The Routledge Handbook of International Political Sociology*. London and New York: Routledge, 193–203.

Coleman, L.M. and Rosenow, D. (2016) Security (studies) and the limits of critique: Why we should think through struggle. *Critical Studies on Security* 4(2): 202–20.

Connolly, W.E. (2017) *Facing the Planetary: Entangled Humanism and the Politics of Swarming*. Durham and London: Duke University Press.

Connolly, W.E. (2011) *A World of Becoming*. Durham and London: Duke University Press.

Connolly, W.E. (2008) *Capitalism and Christianity, American Style*. Durham and London: Duke University Press.

Connolly, W.E. (2002) *The Augustinian Imperative: A Reflection on the Politics of Morality*. New ed. Lanham and Oxford: Rowman & Littlefield Publ. Inc.

Coole, D. and Frost, S. (2010a) Introducing the new materialisms. In: Coole, D. and Frost, S. (eds) *New Materialisms: Ontology, Agency, and Politics*. Durham and London: Duke University Press, 1–43.

Coole, D. and Frost, S. (eds) (2010b) *New Materialisms: Ontology, Agency, and Politics*. Durham and London: Duke University Press.

Cooper, M. (2008) *Life as Surplus: Biotechnology and Capitalism in the Neoliberal Era*. Seattle, WA: University of Washington Press.

Dawkins, R. (1989) [1976] *The Selfish Gene*. Oxford and New York: Oxford University Press.

de Goede, M. and Randalls, S. (2009) Precaution, preemption: Arts and technologies of the actionable future. *Environment and Planning D* 27(5): 859–78.

Deleuze, G. (2004) [1968] *Difference and Repetition*. Transl. Patton, P. London and New York: Continuum.

Deleuze, G. (1999) [1986] *Foucault*. Transl. Hand, S. London: Continuum.

Deleuze, G. (1990) [1969] *The Logic of Sense*. Transl. Lester, M. New York: Columbia University Press.

Deleuze, G. and Guattari, F. (2004) [1980] *A Thousand Plateaus: Capitalism and Schizophrenia*. Transl. Massumi, B. London and New York: Continuum.

Deleuze, G. and Guattari, F. (1994) [1991] *What Is Philosophy?* Transl. Tomlinson, H. and Burchell, G. London: Verso.

Deleuze, G. and Parnet, C. (2006) [1977] *Dialogues II*. Transl. Tomlinson, H. and Habberjam, B. 2nd ed. London and New York: Continuum.

Department of Energy & Climate Change (2015) DEEC public attitudes tracker: Wave 14: Summary of key findings, August. Available at: www.gov.uk/government/uploads/system/uploads/attachment_data/file/450674/PAT_Summary_Wave_14.pdf

Dîaz Pérez, S. (2014) Blow against Monsanto: No GE (or GM) soy allowed in Campeche, Mexico. *Greenpeace Blogpost*, 14 March. Available at: www.greenpeace.org/international/en/news/Blogs/makingwaves/blow-against-monsanto-in-mexico/blog/48524/

Didur, J. and Heffernan, T. (2003) Revisiting the subaltern in the new empire. *Cultural Studies* 17(1): 1–15.

Disch, L. (2008) Representation as 'spokespersonship': Bruno Latour's political theory. *Parallax* 14(3): 88–100.

Dopfer, K. (ed) (2005) *Economics, Evolution and the State: The Governance of Complexity*. Cheltenham and Northampton: Edward Elgar Publishing.

Doty, R.L. (1999) Racism, desire, and the politics of immigration. *Millennium* 28(3): 585–606.

Egorova, Y. and Mantuong, K. (2014) India. In: Macnaghten, P., Carro-Ripalda, S. and Burity, J. (eds) *A New Approach to Governing GM Crops: Global Lessons from the Rising Powers*. Durham, UK: Durham University Working Paper, 67–86.

Egorova, Y., Raina, R.S. and Mantuong, K. (2015) An analysis of the GM crop debate in India. In: Macnaghten, P., Egorova, Y. and Mantuong, K. (eds) *Governing Agricultural Sustainability: Global Lessons from GM Crops*. London and New York: Routledge, 105–35.

Escobar, A. (2007) Worlds and knowledges otherwise: The Latin American modernity/coloniality research program. *Cultural Studies* 21(2–3): 179–210.

Escobar, A. (2004) Beyond the third world: Imperial globality, global coloniality and anti-globalisation social movements. *Third World Quarterly* 25(1): 207–30.

Evans, B. and Reid, J. (2014) *Resilient Life: The Art of Living Dangerously*. Cambridge and Malden: Polity.

Evans, B. and Reid, J. (2013) Dangerously exposed: The life and death of the resilient subject. *Resilience* 1(2): 83–98.

Featherstone, D. (2003) Spatialities of transnational resistance to globalization: The maps of grievance of the Inter-Continental Caravan. *Transactions of the Institute of British Geographers* 28(4): 404–21.

Fontana, M. (2010) Can neoclassical economics handle complexity? The fallacy of the oil spot dynamic. *Journal of Economic Behavior & Organization* 76(3): 584–96.

Fortun, M. (2009) Genes in our knots. In: Atkinson, P., Glasner, P. and Lock, M. (eds) *Handbook of Genetics and Society: Mapping the New Genomic Era*. London and New York: Routledge, 247–59.

Foucault, M. (1991) [1975] *Discipline and Punish: The Birth of the Prison*. Transl. Sheridan, A. London et al.: Penguin.

Foucault, M. and Deleuze, G. (1972) Intellectuals and power: A conversation between Michel Foucault and Gilles Deleuze. Available at: http://libcom.org/library/intellectuals-power-a-conversation-between-michel-foucault-and-gilles-deleuze

Fox Keller, E. (2000) *The Century of the Gene*. Cambridge, MA and London: Harvard University Press.

Fuller, S. (2010) Why science studies has never been critical of science: Some recent lessons on how to be a helpful nuisance and a harmless radical. *Philosophy of the Social Sciences* 30(1): 5–32.

Fuller, S. (2007) *New Frontiers in Science and Technology Studies*. Cambridge and Malden, MA: Polity Press.

Funke, P.N. (2012) The rhizomatic left, neoliberal capitalism and class: Theoretical interventions on contemporary social movements in the global north. *International Critical Thought* 2(1): 30–41.

Glissant, É. (1997) *Poetics of Relation*. Transl. Wing, B. Ann Arbor: The University of Michigan Press.

Hacking, I. (1983) *Representing and Intervening: Introductory Topics in the Philosophy of Natural Science*. Cambridge: Cambridge University Press.

Halvorsen, S. (2012) Beyond the network? Occupy London and the global movement. *Social Movement Studies* 11(3–4): 427–33.

Haraway, D.J. (1997) *Modest-Witness@Second-Millennium: FemaleMan-Meets-Onco-Mouse: Feminism and Technoscience*. New York and London: Routledge.

Hardt, M. and Negri, A. (2011) The fight for 'real democracy' at the heart of Occupy Wall Street. *Foreign Affairs*, 11 October. Available at: www.foreignaffairs.com/articles/136399/michael-hardt-and-antonio-negri/the-fight-for-real-democracy-at-the-heart-of-occupy-wall-street

Hardt, M. and Negri, N. (2009) *Commonwealth*. Cambridge, MA: Belknap Press of Harvard University Press.

Hardt, M. and Negri, A. (2006) *Multitude: War and Democracy in the Age of Empire*. London and New York: Penguin Books.

Hardt, M. and Negri, A. (2000) *Empire*. Cambridge, MA: Harvard University Press.

Harman, G. (2014) *Bruno Latour: Reassembling the Political*. London: Pluto Press.

Herring, R.J. (2009) *Global Rifts over Biotechnology: What Does India's Experience with Bt Cotton Tell Us?* Delhi University, V.T. Krishnamachari Memorial Lecture, 2 December. Available at: http://government.arts.cornell.edu/assets/faculty/docs/herring/KrishnamachariLectureFnlRHv15.pdf

Herring, R.J. (2006) Why did 'Operation Cremate Monsanto' fail? Science and class in India's great terminator-technology hoax. *Critical Asian Studies* 38(4): 467–93.

Herring, R.J. and Rao, N.C. (2012) On the 'failure of Bt cotton': Analysing a decade of experience. *Economics & Political Weekly* 94(18): 45–53.

Hetherington, K. (2013) Beans before the law: Knowledge practices, responsibility, and the Paraguayan soy boom. *Cultural Anthropology* 28(1): 65–85.

Ho, M.-W. (2013) Why GMOs can never be safe. *Institute of Science in Society Report*, 22 July. Available at: www.i-sis.org.uk/Why_GMOs_Can_Never_be_Safe.php

Ho, M.-W. (2011) Preface: Beauty and truth in science and art. In: *Celebrating ISIS: Quantum Jazz Biology Medicine Art: Commemorating Institute of Science in Society (1999–2011), 26–27 March*. London: Institute of Science in Society, 9–12.

Ho, M.-W. (2010) Development and evolution revisited. In: Hood, K. et al. (eds) *Developmental Science, Behavior, and Genetics*. Malden, Oxford and Chichester: Blackwell Publishing, 61–109.

Ho, M.-W. (2003) *Living with the Fluid Genome*. London and Penang: Institute of Science in Society and Third World Network.

Ho, M.-W. (2000) The entangled universe. *Yes! Magazine*, 31 March. Available at: www.yesmagazine.org/issues/new-stories/329

Ho, M.-W. (1998) *Genetic Engineering: Dream or Nightmare? The Brave New World of Bad Science and Big Business*. Bath: Gateway.

Ho, M.-W. (1993) *The Rainbow and the Worm: The Physics of Organisms*. Singapore and London: World Scientific Publishing.

Hood, K.E. et al. (2010) Developmental systems, nature-nurture, and the role of genes in behavior and development: On the legacy of Gilbert Gottlieb. In: Hood, K.E. et al. (eds) *Developmental Science, Behavior, and Genetics*. Malden, Oxford and Chichester: Blackwell Publishing, 3–12.

Institute for Science in Society Report (2004) *Death of the Central Dogma*, 3 September. Available at: http://www.i-sis.org.uk/DCD.php

Jasanoff, S. (2005) *Designs on Nature: Science and Democracy in Europe and the United States*. 4th printing, 2007. Princeton and Woodstock: Princeton University Press.

Jordan, W. (2014) Many in Britain still sceptical of GM foods. *YouGov*, 21 February. Available at: https://yougov.co.uk/news/2014/02/21/many-britain-remain-sceptical-gm-foods/

Kauffman, S. (1996) *At Home in the Universe: The Search for Laws of Self-Organization and Complexity*. London et al: Penguin Books.

Kaviraj, S. (2011) On the enchantment of the state: Indian thought on the role of the state in the narrative of modernity. In: Gupta, A. and Sivaramakrishnan, K. (eds) *The State in India after Liberalization: Interdisciplinary Perspectives*. London and New York: Routledge, 31–48.

Kerber, W. (2005) Applying evolutionary economics to public policy: The example of competitive federalism in the EU. In: Dopfer, K. (ed) *Economics, Evolution and the State: The Governance of Complexity*. Cheltenham and Northampton: Edward Elgar Publishing, 296–324.

Kinchy, A. (2012) *Seeds, Science, and Struggle: The Global Politics of Transgenic Crops*. Cambridge, MA: MIT Press.

Kousis, M. (2010) New challenges for twenty-first century environmental movements: Agricultural biotechnology and nanotechnology. In: Redclift, M.R. and Woodgate, G. (eds) *The International Handbook of Environmental Sociology*. Cheltenham and Northampton, MA: Edward Elgar Publishing, 226–44.

Latour, B. (2017) Why Gaia is not a God of totality. *Theory, Culture & Society* 34(2–3): 61–81.

Latour, B. (2013) *An Inquiry into Modes of Existence: An Anthropology of the Moderns*. Transl. Porter, C. Cambridge, MA and London: Harvard University Press.

Latour, B. (2005) *Reassembling the Social: An Introduction to Actor-Network-Theory*. Oxford et al.: Oxford University Press.

Latour, B. (2004) *Politics of Nature: How to Bring the Sciences into Democracy*. Transl. Porter, C. Cambridge, MA and London: Harvard University Press.

Latour, B. (1999) Discussion: For David Bloor . . . and beyond: A reply to David Bloor's 'Anti-Latour'. *Studies in History and Philosophy of Science* 30(1): 113–29.

Latour, B. (1983) Give me a laboratory and I will raise the world. In: Knorr-Cetina, K.D. and Mulkay, M. (eds) *Science Observed: Perspectives on the Study of Science*. London: Sage Publications, 141–70.

Leach, D.K. (2013) Culture and the structure of tyrannylessness. *The Sociological Quarterly* 54: 159–228.

Leadsom, A. (2016) Realising the vision for a new fleet of nuclear power stations. *Speech at the 8th Nuclear New Build Forum*, 20 April. Available at: www.gov.uk/government/speeches/realising-the-vision-for-a-new-fleet-of-nuclear-power-stations

Lezaun, J. (2006) Creating a new object of government: Making genetically modified organisms traceable. *Social Studies of Science* 36(4): 499–531.

Lovelock, J. (1990) *The Ages of Gaia: A Biography of the Living Earth*. New York and London: Bantam.

Lugones, M. (2010) Toward a decolonial feminism. *Hypatia* 25(4): 742–59.

Lugones, M. (2007) Heterosexualism and the colonial/modern gender system. *Hypathia* 22(1): 186–219.

Lugones, M. (2003) *Pilgrimages/Peregrinajes: Theorizing Coalition Against Multiple Oppressions*. Lanham: Rowman & Littlefield Publishers, Inc.

Lynas, M. (2015) How I got converted to G.M.O. food. *The New York Times*, 24 April. Available at: www.nytimes.com/2015/04/25/opinion/sunday/how-i-got-converted-to-gmo-food.html?_r=2

Macnaghten, P., Carro-Ripalda, S. and Burity, J. (2015) Researching GM crops in a global context. In: Macnaghten, P., Egorova, Y. and Mantuong, K. (eds) *Governing Agricultural Sustainability: Global Lessons from GM Crops*. London and New York: Routledge, 5–32.

Macnaghten, P., Egorova, Y. and Mantuong, K. (eds) (2015) *Governing Agricultural Sustainability: Global Lessons from GM Crops*. London and New York: Routledge.

Madsen, S.T. (2001) The view from Vevey. *Economic & Political Weekly* 36(39): 3733–42.

Maldonado-Torres, N. (2007) On the coloniality of Being: Contributions to the development of a concept. *Cultural Studies* 21(2–3): 240–70.

McDonald, K. (2006) *Global Movements: Action and Culture*. Oxford: Blackwell.

McWeeny, J. (2014) Topographies of flesh: Women, nonhuman animals, and the embodiment of connection and difference. *Hypatia* 29(2): 269–86.

Meacher, M. (2011) Foreword. In: *Celebrating ISIS: Quantum Jazz Biology Medicine Art: Commemorating Institute of Science in Society (1999–2011), 26–27 March*. London: Institute of Science in Society, 8.

Means, R. (1980) For America to live, Europe must die. *Speech given at the Black Hills International Survival Gathering*, July. Available at: https://endofcapitalism.com/2010/10/17/revolution-and-american-indians-marxism-is-as-alien-to-my-culture-as-capitalism/

Mignolo, W.D. (2011) Epistemic disobedience and the decolonial option: A manifesto. *Transmodernity* 1(2): 44–66.

Mignolo, W.D. (2010) Introduction: Coloniality of power and de-colonial thinking. In: Mignolo, W.D. and Escobar, A. (eds) *Globalization and the Decolonial Option*. London and New York: Routledge, 1–21.

Mignolo, W.D. (2009) Epistemic disobedience, independent thought and de-colonial freedom. *Theory, Culture & Society* 26(7–8): 1–23.

Mignolo, W.D. (2007) Delinking: The rhetoric of modernity, the logic of coloniality and the grammar of de-coloniality. *Cultural Studies* 21(2–3): 449–514.

Mignolo, W.D. (2002) The geopolitics of knowledge and the colonial difference. *South Atlantic Quarterly* 101(1): 57–96.

Mignolo, W.D. (2000) *Local Histories/Global Designs: Coloniality, Subaltern Knowledges, and Border Thinking*. Princeton, NJ: Princeton University Press.

Monbiot, G. (2013) Nuclear scare stories are a gift to the truly lethal coal industry. *The Guardian*, 16 December. Available at: www.theguardian.com/commentisfree/2013/dec/16/nuclear-scare-stories-coal-industry

Moraga, C. (1983) *Loving in the War Years: Lo que nunca paso por sus labios*. Boston: South End Press.

Nagaraj, D.R. (2012) [1996] Anxious Hindu and angry farmer: Notes on the culture and politics of two responses to globalization in India. In: Nagaraj, D.R. *Listening to the Loom: Essays on Literature, Politics and Violence*. Ed. Shobhi, P.D.C. Ranikhet: Permanent Black, 284–307.

Nanjundaswamy, M.D. (2003) Cremating Monsanto: Genetically modified fields on fire. In: Notes from Nowhere (ed) *We Are Everywhere: The Irresistible Rise of Global Anticapitalism*. London and New York: Verso, 152–9.

Nanjundaswamy, M.D. (1999) To establish people's power. *Statement at the Demonstration against the World Economic Summit*, Cologne, 19 June. Available at: http://caravan.squat.net/ICC-en/statement.htm

Neumann-Held, E.M. and Rehmann-Sutter, C. (2006) Introduction. In: Neumann-Held, E.M. and Rehmann-Sutter, C. (eds) *Genes in Development: Re-Reading the Molecular Paradigm*. Durham, NC: Duke University Press, 1–11.

Omvedt, G. (1998) Terminating choice. *The Hindu*, 14 December. Accessed via *Factiva*, https://www.dowjones.com/products/factiva/

O'Neill, K. (2004) Transnational protest: States, circuses, and conflict at the frontline of global politics. *International Studies Review* 6(2): 233–51.

Open Letter from World Scientists to All Governments (2000) 1 September. Available at: www.i-sis.org.uk/list.php

Ortega, M. (2001) 'New Mestijas,' ' "World"-Travelers,' and '*Dasein*': Phenomenology and the multi-voiced, multi-cultural self. *Hypatia* 16(3): 1–29.

Oyama, S. (2000) *The Ontogeny of Information: Developmental Systems and Evolution*. Rev. and enl. ed. Durham, NC: Duke University Press.

Pellizzoni, L. (2011) Governing through disorder: Neoliberal environmental governance and social theory. *Global Environmental Change* 21(3): 795–803.

Pleyers, G. (2010) *Alter-Globalization: Becoming Actors in a Global Age*. Cambridge and Malden: Polity Press.

Poulter, S. (2015) Senior academic condemns 'deluded' supporters of GM food as being 'anti-science' and ignoring evidence of dangers. *Mail Online*, 4 March. Available at: www.dailymail.co.uk/news/article-2979645/Senior-academic-condemns-deluded-supporters-GM-food-anti-science-ignoring-evidence-dangers.html

Prentoulis, M. and Thomassen, L. (2013) Political theory in the square: Protest, representation and subjectification. *Contemporary Political Theory* 12(3): 166–84.

Pretty, J. (1998) Feeding the world? *Splice* 4(6). Available at: ngin.tripod.com/article2.htm

Quijano, A. (2007) Coloniality and modernity/rationality. *Cultural Studies* 21(2–3): 168–78.

Rancière, J. (2010) *Dissensus: On Politics and Aesthetics*. Ed. and Transl. Corcoran, S. London and New York: Continuum.

Rancière, J. (1999) *Disagreement: Politics and Philosophy*. Minneapolis: University of Minnesota Press.

Reid, J. (2012) The disastrous and politically debased subject of resilience. *Development Dialogue* 58(April): 67–79.

Reid, J. (2011) The vulnerable subject of liberal war. *South Atlantic Quarterly* 110(3): 770–9.

Reitan, R. and Gibson, S. (2012) Climate change or social change? Environmental and leftist praxis and participatory action research. *Globalizations* 9(3): 395–410.

Robinson, C. (2015) What BBC's Panorama got wrong on GMOs. *GM Watch*, 9 June. Available at: www.gmwatch.org/en/news/latest-news/16221-what-bbc-s-panorama-got-wrong-on-gmos

Rootes, C. (2012) New issues, new forms of action? Climate change and environmental activism in Britain. In: van Deth, J.W. and Maloney, W. (eds) *New Participatory Dimensions in Civil Society: Professionalization and Individualized Collective Action*. London: Routledge, 46–68.

Rosenow, D. (2017) Flucht vor "herrschender" Kritik? Occupy Wall Street und Affektpolitik. In: Daase, C. et al. (eds) *Herrschaft in den International Beziehungen*. Wiesbaden: Springer VS, 201–22.

Rosenow, D. (2013) Nomadic life's counter-attack: Moving beyond the subaltern's voice. *Review of International Studies* 39(2): 415–33.

Rosenow, D. (2012) Dancing life into being: Genetics, resilience and the challenge of complexity theory. *Security Dialogue* 43(6): 531–47.

Rosenow, D. (2009) Decentring global power: The merits of a Foucauldian approach to international relations. *Global Society* 23(4): 497–517.

Routledge, P. (2008) Acting in the network: ANT and the politics of generating associations. *Environment and Planning D* 26(2): 199–217.

Routledge, P., Cumbers, A. and Nativel, C. (2006) Entangled logics and grassroots imaginaries of global justice networks. *Environmental Politics* 15: 839–59.

Roy, D., Herring, R.J. and Geisler, C.C. (2007) Naturalising transgenics: Official seeds, loose seeds, and risk in the decision matrix of Gujarati cotton farmers. *Journal of Development Studies* 43(1): 158–76.

Samaddar, R. (2010) *Emergence of the Political Subject*. New Delhi et al.: Sage Publications.

Samaddar, R. (2007) *The Materiality of Politics: The Technologies of Rule*, Volume 1. London, New York and Delhi: Anthem Press.

Sandoval, C. (1998) Mestizaje as method: Feminists-of-color challenge the canon. In: Trujillo, C. (ed) *Living Chicana Theory*. Berkeley: Third Woman Press, 352–70.

Sandoval, C. (1995) Feminist forms of agency and oppositional consciousness: U.S. third world feminist criticism. In: Gardiner, J.K. (ed) *Provoking Agents: Gender and Agency in Theory and Practice*. Urbana: University of Illinois Press, 208–19.

Sarkar, S. (2006) From genes as determinants to DNA as resource: Historical notes on development and genetics. In: Neumann-Held, E.M. and Rehmann-Sutter, C. (eds) *Genes in Development: Re-Reading the Molecular Paradigm*. Durham, NC: Duke University Press, 77–95.

Schnurr, M.A. and Mujabi-Mujuzi, S. (2014) 'No one asks for a meat they've never eaten': Or, do African farmers want genetically modified crops? *Agriculture and Human Values* 31(4): 643–8.

Schurman, R. and Munro, W. (2010) *Fighting for the Future of Food: Activists vs. Agrobusiness in the Struggle over Biotechnology*. Minneapolis: Minnesota University Press.

Schurman, R. and Munro, W. (2009) Targeting capital: A cultural economy approach to understanding the efficacy of two anti-genetic engineering movements. *American Journal of Sociology* 15(1): 155–202.

Shah, E. (2008) What makes crop biotechnology find its roots? The technological culture of Bt cotton in Gujarat, India. *The European Journal of Development Research* 20(3): 432–47.

Shiva, V. (2015) Pests, pesticides and propaganda: The story of Bt cotton. 10 October. Available at: http://vandanashiva.com/?p=317

Shiva, V. (2009) Woman and the gendered politics of food. *Philosophical Topics* 37(2): 17–32.

Shiva, V. (2007) Not so green revolution: Lessons from India. In: Nærstad, A. (ed) *Africa Can Feed Itself*. Oslo: Organizing committee of the conference in Oslo, Norway, 6–8 June 2007, 142–50.

Shiva, V. (2006) The British in partnership with the developing world. In: Fabian Society (ed) *2025: What Next for the Make Poverty History Generation?* London: Fabian Society, 22–5.

Shiva, V. (2000) *Seeds of Suicide: The Ecological and Human Costs of Globalization of Agriculture*. Delhi: Research Foundation for Science, Technology and Ecology.

Shiva, V. (1989) *Staying Alive: Women, Ecology and Development*. London: Zed Books.

Shiva, V., Emani, A. and Jafri, A.H. (1999) Globalisation and threat to seed security: Case of transgenic cotton trials in India. *Economic & Political Weekly* 34(10–11): 601–13.

Shiva, V. et al. (eds) (2000) *Licence to Kill*. New Delhi: Research Foundation for Science, Technology and Ecology.

Sinha, S. (2008) Lineages of the developmentalist state: Transnationality and village India, 1900–1965. *Comparative Studies in Society and History* 50(1): 57–90.

Sinha, S. (2003) Development counter-narratives: Taking social movements seriously. In: Sivaramakrishnan, K. and Agrawal, A. (eds) *Regional Modernities: The Cultural Politics of Development in India*. Stanford: Stanford University Press, 286–312.

Sivaramakrishnan, K. and Agrawal, A. (2003) Regional modernities in stories and practices of development. In: Sivaramakrishnan, K. and Agrawal, A. (eds) *Regional Modernities: The Cultural Politics of Development in India*. Stanford: Stanford University Press, 41–53.

Smith, N. (2000) Global seattle. *Environment and Planning D* 18: 1–3.

Social Movements Assembly (2013) Declaration of the social movements assembly of the world social forum, 31 March. Available at: http://rio20.net/en/iniciativas/declaration-of-the-social-movements-assembly-of-the-world-social-forum-2013/

Spivak, G.C. (2004) Righting wrongs. *The South Atlantic Quarterly* 103(2–3): 523–81.

Spivak, G.C. (1999) *A Critique of Postcolonial Reason: Toward a History of the Vanishing Present*. Cambridge, MA and London: Harvard University Press.

Spivak, G.C. (1988) Can the subaltern speak? In: Nelson, C. and Grossberg, L. (eds) *Marxism and the Interpretation of Culture*. Urbana and Chicago: University of Illinois Press, 271–313.

Steger, M.B., Goodman, J. and Wilson, E.K. (2013) *Justice Globalism: Ideology, Crises, Policy*. London et al.: Sage Publications.

Steger, M.B. and Wilson, E.K. (2012) Anti-globalization or alter-globalization? Mapping the political ideology of the global justice movement. *International Studies Quarterly* 56(3): 439–54.

Stengers, I. (2010) Including nonhumans in political theory: Opening Pandora's box? In: Brown, B. and Whatmore, S. (eds) *Political Matter: Technoscience, Democracy, and Public Life*. Minneapolis: University of Minnesota Press, 3–34.

Stone, G.D. (2015) Biotechnology, schismogenesis, and the demise of uncertainty. *Washington University Journal of Law & Policy* 47(1): 29–49.

Stone, G.D. (2012) Constructing facts: Bt cotton narratives in India. *Economic & Political Weekly* 47(38): 62–70.

Stone, G.D. (2007) Agricultural deskilling and the spread of genetically modified cotton in Warangal. *Current Anthropology* 48(1): 67–103.

Stone, G.D. and Flachs, A. (2014) The problem with the farmer's voice. *Agriculture and Human Values* 31(4): 649–53.

Stotz, K. (2009) Experimental philosophy of biology: Notes from the field. *Studies in History and Philosophy of Science* 40: 233–7.

Strathausen, C. (2010) Epistemological reflections on minor points in Deleuze. *Theory & Event* 13(4). Available at: https://muse.jhu.edu/article/407139/summary

Suárez-Krabbe, J. (2016) *Race, Rights and Rebels: Alternatives to Human Rights and Development from the Global South*. London: Rowman & Littlefield International.

Sullivan, S. (2005) An *other* world is possible? On representations, rationalism and romanticism in social forums. *Ephemera* 5(2): 370–92.

Swyngedouw, E. (2007) Impossible 'sustainability' and the postpolitical condition. In: Krueger, R. and Gibbs, D.C. (eds) *The Sustainability Paradox: Urban Political Economy in the United States and Europe*. New York: Guilford Press, 13–40.

Tharoor, S. (2017) *Inglorious Empire: What the British Did to India*. London: Hurst & Co.

Tilley, L. (2017) Resisting piratic method by doing research otherwise. *Sociology* 51(1): 27–42.

Todd, Z. (2016) An indigenous feminist's take on the ontological turn: 'Ontology' is just another word for colonialism. *Journal of Historical Sociology* 29(1): 4–22.

Tolia-Kelly, D.P. (2006) Affect: An ethnocentric encounter? Exploring the 'universalist' imperative of emotional/affectual geographies. *Area* 38(2): 213–17.

Torgersen, H. et al. (2002) Promise, problems and proxies: Twenty-five years of debate and regulation in Europe. In: Bauer, M.W. and Gaskell (eds) *Biotechnology: The Making of a Global Controversy*. Cambridge: Cambridge University Press, 21–94.

Tormey, S. (2012) Occupy Wall Street: From representation to post-representation. *Journal of Critical Globalisation Studies* 5: 32–7.

Turner, V. (1974) *Dramas, Fields and Metaphors: Symbolic Action in Human Society*. Ithaca and London: Cornell University Press.

Visvanathan, S. and Parmar, C. (2003) A biotechnology story: Notes from India. *Economic & Political Weekly* 37(27): 2714–24.

Viveiros de Castro, E. (2015) *The Relative Native: Essays on Indigenous Conceptual Worlds*. Chicago: Hau Books.

von Weizsäcker, C. (2005) Is the notion of progress compatible with an evolutionary view of the economy? In: Dopfer, K. (ed) *Economics, Evolution and the State: The Governance of Complexity*. Cheltenham and Northampton: Edward Elgar Publishing, 43–57.

Walker, J. and Cooper, M. (2011) Genealogies of resilience: From systems ecology to the political economy of crisis adaption. *Security Dialogue* 14(2): 143–60.

Watson, G. (2015) GM, PR & the BBC. *Riverford Newsletter*, 22 June.

Watson, J. (2004) *DNA: The Secret of Life*. London: Arrow Books.

Whatmore, S. (2009) Mapping knowledge controversies: Science, democracy and the redistribution of expertise. *Progress in Human Geography* 33(5): 587–98.

Whatmore, S. (2002) *Hybrid Geographies: Natures, Cultures, Spaces*. Thousand Oaks, London and New Delhi: Sage Publications.

White, S.K. (2000) *Sustaining Affirmation: The Strengths of Weak Ontology in Political Theory*. Princeton and Oxford: Princeton University Press.

Wilkin, P. (2003) Against global governance? Tracing the lineage of the Anti-Globalisation Movement. In: Cochrane, F., Duffy, R. and Selby J. (eds) *Global Governance, Conflict and Resistance*. London: Palgrave MacMillan, 78-96.

Williams, J. (2005) *The Transversal Thought of Gilles Deleuze: Encounters and Influences*. Manchester: Clinamen Press.

Williams, R.B.H. and Luo, O.J. (2010) Complexity, post-genomic biology and gene expression programs. In: Dewar, R.L. and Detering, F. (eds) *Complex Physical, Biophysical and Ecophysical Systems*. Singapore: World Scientific Publishing, 319–52.

World Trade Organization (2006) European communities: Measures affecting the approval and marketing of biotech products: Report of the panel. Available at: www.wto.org/english/tratop_e/dispu_e/cases_e/ds291_e.htm

Worth, O. and Buckley, K. (2009) The world social forum: Postmodern prince or court jester? *Third World Quarterly* 30(4): 649–61.

The Wretched of the Earth bloc (2016) Open letter from the Wretched of the Earth bloc to the organisers of the People's Climate March of Justice and Jobs, 16 December. Available at: https://reclaimthepower.org.uk/news/open-letter-from-wretched-of-the-earth-bloc-to-organisers-of-peoples-climate-march/

Wynter, S. (2003) Unsettling the coloniality of being/power/truth/freedom: Towards the human, after man, its overrepresentation: An argument. *The New Centennial Review* 3(3): 257–337.

Žižek, S. (1999) *The Ticklish Subject: The Absent Centre of Political Ontology*. London: Verso.

Index

Milton Keynes UK
Ingram Content Group UK Ltd.
UKHW040051071024
449327UK00019B/490